# CHESAPEAKE BAY
## Crabs

# CHESAPEAKE BAY

## Crabs

### JUDY COLBERT

**PELICAN PUBLISHING COMPANY**
GRETNA 2011

Copyright © 2011
By Judy Colbert
All rights reserved

*The word "Pelican" and the depiction of a pelican are trademarks of Pelican Publishing Company, Inc., and are registered in the U.S. Patent and Trademark Office.*

**Library of Congress Cataloging-in-Publication Data**

Colbert, Judy.
  Chesapeake Bay crabs / Judy Colbert.
    p. cm.
  Includes index.
  ISBN 978-1-58980-974-1 (hardcover : alk. paper) — ISBN 978-1-4556-1504-9 (e-book)  1.  Cooking (Crabs) 2.  Crabs—Chesapeake Bay Region (Md. and Va.) 3.  Blue crab. 4.  Cookbooks.  I. Title.

TX754.C83C65 2011
641.6'95—dc23

                                      2011012270

*All images courtesy Judy Colbert unless otherwise noted.*

Printed in China
Published by Pelican Publishing Company, Inc.
1000 Burmaster Street, Gretna, Louisiana 70053

"You cannot teach a crab to walk straight."

Aristophanes

"Thank goodness."

Judy Colbert

Thanks to:

Each and every person who contributed a recipe, an idea, a photograph, a concept, and knowledge about the Chesapeake Bay, its watermen, and its blue crabs.

Joyce A. Stinnett Baki, Calvert County Department of Economic Development, and Aubrey Manzo, Southern Delaware Tourism, who gave so much time and provided so much assistance.

Jack Brooks at J. M. Clayton Co. and Mike and Don Storm at Shoreline Seafood.

# Contents

*Part I: Crabs*

    Basic Crab Information . . . . . . . . . . . . . . . . . . . . . . . . . . . 13

    Bay Explorations . . . . . . . . . . . . . . . . . . . . . . . . . . . . . . . . . 28

    Bon Voyage . . . . . . . . . . . . . . . . . . . . . . . . . . . . . . . . . . . . . 34

    Bernie Fowler . . . . . . . . . . . . . . . . . . . . . . . . . . . . . . . . . . . 36

    Crab-u-lous Time . . . . . . . . . . . . . . . . . . . . . . . . . . . . . . . . 40

    Buying Crabs . . . . . . . . . . . . . . . . . . . . . . . . . . . . . . . . . . . 47

    Catching Crabs . . . . . . . . . . . . . . . . . . . . . . . . . . . . . . . . . 51

    Cooking Crabs . . . . . . . . . . . . . . . . . . . . . . . . . . . . . . . . . . 53

    Picking Steamed Crabs . . . . . . . . . . . . . . . . . . . . . . . . . . . . 60

    Soft-Shell Crabs . . . . . . . . . . . . . . . . . . . . . . . . . . . . . . . . . 62

    Glossary . . . . . . . . . . . . . . . . . . . . . . . . . . . . . . . . . . . . . . . 63

    Food Tips . . . . . . . . . . . . . . . . . . . . . . . . . . . . . . . . . . . . . . 67

    Blue Crab Festivals . . . . . . . . . . . . . . . . . . . . . . . . . . . . . . . 74

*Part II: Recipes*

    Breakfast . . . . . . . . . . . . . . . . . . . . . . . . . . . . . . . . . . . . . . 81

    Appetizer and Dips . . . . . . . . . . . . . . . . . . . . . . . . . . . . . . . 93

    Crab Cakes . . . . . . . . . . . . . . . . . . . . . . . . . . . . . . . . . . . . 131

    Salads and Sandwiches . . . . . . . . . . . . . . . . . . . . . . . . . . . 163

    Soups . . . . . . . . . . . . . . . . . . . . . . . . . . . . . . . . . . . . . . . . 179

    Sauces . . . . . . . . . . . . . . . . . . . . . . . . . . . . . . . . . . . . . . . 203

    Entrées . . . . . . . . . . . . . . . . . . . . . . . . . . . . . . . . . . . . . . . 207

    Desserts . . . . . . . . . . . . . . . . . . . . . . . . . . . . . . . . . . . . . . 257

    Index . . . . . . . . . . . . . . . . . . . . . . . . . . . . . . . . . . . . . . . . 265

# CHESAPEAKE BAY
## Crabs

# Part I: Crabs

# Basic Crab Information

Welcome to the beautiful Chesapeake Bay, the Land of Pleasant Living, and its prized catch, the blue crab (*Callinectes sapidus*, meaning savory beautiful swimmer), a crustacean (along with lobsters and shrimp). One could say the Bay area is crab-u-lous.

The Bay is the largest estuary in the United States (there are 840 other estuaries in the country), measuring about two hundred miles north to south and between almost three miles wide at its northern end to thirty miles wide where the Potomac River feeds into the Bay on its way to meet the Atlantic Ocean. It drains 150 rivers and streams from Delaware, Maryland, New York, Pennsylvania, Virginia, West Virginia, and Washington, D.C., creating a basin of more than 64,000 square miles. The largest rivers flowing into the Bay are the Susquehanna, Patapsco, Chester, Choptank, Patuxent, Nanticoke, Potomac, and Pocomoke feeding in from Maryland and the Rappahannock, York, and James rivers feeding in from Virginia.

The bay yields more fish and shellfish (about 45,000 short tons) than any other estuary in the United States.

Early Maryland settlers lived within close proximity of the Bay, more as a transportation route than because of the food source contained within its waters. The settlers worked the land as their primary nourishment supply. It was plentiful, inexpensive, and nutrient-rich. Anyone fishing in the Bay's waters was as likely to be doing so for a recreational as for commercial purpose. The equipment available for catching fish was pretty much limited to hook and line and raking the shore for oysters.

With a rapidly increasing population, hurricanes, weather changes, and various wars and skirmishes, fishing (and crabbing) was not looked upon as a reliable food supply or a steady income. On the other hand, with the development of rail shipping and refrigerated transportation, the ability to move Bay products beyond its shorelines started increasing. The availability of canning equipment beginning in the late 1870s also increased the ability to transport crabmeat to places beyond the local communities.

Virginia's watermen have been catching blue crabs and eighty-six other varieties of seafood since the first settlement in Jamestown. Today, the state is the nation's fourth largest seafood producer and the largest on the East Coast.

Capt. John Smith (1580-1631) is credited as being the first European to explore the Bay, in 1607 and 1608, during his voyages through the waterway seeking a Northwest Passage to the western states. Fortunately, he made meticulous notes and his maps of the Bay, published in 1612, served as the most important point of information for Bay settlers. They have also been instrumental historically for those who are working to restore the Bay to a more viable condition.

The Captain John Smith Chesapeake National Historic Trail, tracing 3,000 miles of historic voyages by Smith, is the first national water trail in the country. Through the trail, you can learn about the importance of and dangers to the Bay, the maritime history, the English settlement of this part of the country, and the traditions of the Native Americans who lived here.

Friends of the John Smith Chesapeake Trail, the Chesapeake Bay Foundation, Sultana Projects, Inc., National Oceanic & Atmospheric Administration, Verizon Wireless, National Geographic Society, and the Conservation Fund have been working together on this trail project.

Sultana's John Smith 400 Project built a replica of Smith's twenty-eight-foot shallop (a type of boat) that was crewed by twelve modern explorers, naturalists, educators, and historians who duplicated Smith's travels on 1,500 miles of water under sail and oar power in the summer of 2007. The 121-day trip stopped at more than twenty ports along the way, taking the voyage to the public.

Even without a boat, you can follow his travels through the Captain John Smith Chesapeake National Historic Water Trail. The Chesapeake Bay Interpretive Buoy System (CBIBS) allows you to explore his voyages vicariously through the Internet. Each buoy marked on the map has an audio presentation about the history, geography, and a seasonal report. The buoys are located at Susquehanna, Patapsco, Annapolis (Severn River), upper and lower Potomac River, Gooses Reef, Stingray Point, James River at Jamestown, and Elizabeth River at Norfolk.

In 1917 Maryland implemented its first measures against the harvest of sponge crabs and Virginia followed suit in the next decade. Virginia lifted the ban in 1932, and while the market lessened during the Great Depression and Prohibition, the development of the crab pot in the mid-1930s started it back upwards again. Although the demand for crabmeat lessened during World War II, it was not a rationed food so it was fairly readily available. The industry benefited in postwar days because of the new products and equipment developed during the war.

Blue crabs are vital to the Bay because they provide a food source for humans and other Bay inhabitants (other crabs, striped bass, catfish, and other species dine on crabs—particularly soft-shelled and juveniles), employment (catching them; cooking, picking, packing, and shipping crabmeat; preparing crab for consumption, etc.), and are a predator that feeds on bivalves, fish, plants, and other organisms that filter down to the bottom of the Bay where they live.

Although the blue crab of the Chesapeake has a limited season (April 1 through December 15), it's estimated that the Bay provides one-third of the crabs consumed in this country. Other blue crabs and similar relatives come from the Gulf of Mexico and Asian waters. The crab is among an estimated three hundred species of fish and shellfish, with the striped bass (rockfish) being the state fish. It had been fished nearly to extinction until a moratorium was enacted so the population could increase. The catch is still strictly limited and fishermen are regularly charged with violating the number, size, and weight limits. Commercially caught and sold striped bass must be checked in to verify where and how each fish was caught and must have colored tags indicating where they were caught.

Starting in 2008, Maryland and Virginia joined forces to impose strict limits on the number of female crabs that could be caught. The aim of the program is to increase the crab population so that the life and business of the commercial crabbers could continue to be worthwhile and profitable, and to assure the availability of crabs for leisure crabbers and diners around the Bay and the world. They also prohibited the issuance of new commercial licenses.

Egg-bearing females (sooks or spongys) that have mated with Jimmies during the summer start migrating to the warmer waters of the southern Bay in mid-September. Although the outcome has been spectacular and one might think it beneficial to maintain the ban for

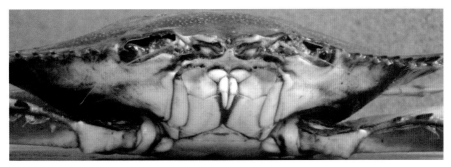

*Chesapeake Bay blue crab*

at least another year or two, the crabbers keep asking to be allowed a larger catch.

The results, says the Chesapeake Bay Program, indicate a 70 percent increase in adult crabs, the first major jump since 1993; however, the number of juveniles is still below the desired numbers to sustain this industry.

According to Virginia governor Bob McDonnell and Maryland governor Martin O'Malley, the blue crab population has increased the last two years because of this stock-rebuilding program. The 2010 survey estimated the population had increased by 60 percent, to 658 million crabs (give or take a dozen or two), the largest number since 1997.

This in no way indicates the Bay is as healthy as it should or could be, but it is a sign of improvement. Fertilizers and other chemicals used in farming are a major source of pollution. Run-off from acres and acres of parking lots, highways and streets, commercial properties, and housing developments is another major source.

One slightly unusual cause of pollution is referred to as derelict crab traps. These are traps that have come loose from crab lines laid by commercial crabbers. They are considered an eye sore, a navigational hazard, and a danger to other species that become caught accidentally in the traps. They are considered derelict or ghost traps, and they keep company with fishing gear, beverage bottles and cans, items that fall off boats and piers, and other marine debris that manage to break away from where it belongs. No one is saying crabbers deliberately drop the traps, for they cost about $20.00 to $30.00 each. Quite often it's pleasure boats that sail or motor by that sever the lines and cause the traps to sink.

The concern is about "any manmade solid material that is—

*Crab traps, during winter.*

directly or indirectly, intentionally or unintentionally—disposed of or abandoned into the marine environment. It may end up in the water or on the shoreline from people on boats or along the shoreline, or can be washed out to sea via rivers, streams, and storm drains. Because marine debris can hurt the ecosystem, NOAA (National Oceanographic and Atmospheric Administration) is researching what those impacts are—and how to mitigate those negative effects."

The NOAA Chesapeake Bay office studied the effects of this gear on the habitat to determine the "safety, nuisance, environmental, and economic impacts in coastal waters," noting that the Bay's "crab fishery—the nation's largest—uses metal traps as the primary method of harvest."

An estimated "300,000 crab traps are deployed each summer day and from that number commercial fishermen may lose from 20 to 30 percent of their traps for a variety of reasons." Although each trap is attached to a marker buoy that identifies its owner and helps with retrieval, the traps become detached when a vessel's propeller cuts a line or wave action and current otherwise sever it. The traps can't be located or retrieved and any catch can't be harvested. These traps may trap, wound, and eventually kill "crabs, fish, birds, reptiles, and aquatic mammals; degrading marine ecosystems and sensitive habitats . . . and forming hazards to recreational, commercial, and military navigation."

The ghost traps, although empty of set bait, can attract crabs, fish, and terrapins, which then cannot escape; they die and become bait that attracts other sea life and the cycle continues. Recreational fishers are also concerned because this cycle has an impact on their catch. One estimate noted that each pot or trap could attract about four dozen crabs or nearly a bushel, and when that's multiplied times 17,000 or more traps, a lot of the potential crab catch is lost.

The NOAA Chesapeake Bay Office, the Maryland Department of Natural Resources, Virginia Marine Resources Commission, Virginia Institute of Marine Science, and the NOAA Marine Debris Program surveyed the Bay from late February through late March 2007 (harvesting was closed to commercial fishery, which assured that any traps located were derelict).

These studies indicated about 85,000 crab traps were on the bottom of the Maryland portion of the Bay and another 35,000 in the Virginia portion of the Bay and its tributaries. One option for future elimination or prevention of this problem is to include "biodegradable components into crab trap design and construction."

To help eliminate or correct the problem, the Virginia Marine Resources Commission and the Virginia Institute of Marine Science employed watermen who would have worked in the 2008-09 winter dredge season (which was closed by the Virginia Marine Resources Commission) to instead retrieve derelict traps in a structured and environmentally sensitive manner. During the two-year program, watermen found 8,800 crab and eel pots the first year and about as many the second year. Maryland's Department of Natural Resources conducted a similar program from February 22 to April 1, 2010.

In a short time, using grappling hooks dragged across the bottom of the Bay and the West and Rhode rivers off Maryland's Anne Arundel County coast and other rivers, more than 9,000 crab pots, eel pots, nets, and other debris were collected. It's not a lot, but it's a start. The watermen who trawled for these traps were required to log where each pot was recovered, note what condition it was in, and whether there was any fish or crabs in the trap.

These derelict traps are not a Chesapeake Bay exclusive. A removal program in Mississippi in 2010 caught 347 traps, which were recycled.

Although Maryland surrounds most of the Bay, Virginia plays an important part in its health. According to the Virginia Institute of Marine Science, the state's seafood industry is one of the oldest industries in the United States and one of the Commonwealth's largest, with an annual economic impact of more than $500 million. Blue crabs represent the second largest economic value, after sea

*King crab claws*

scallops and before striped bass, summer flounder, croaker, spot, and clams.

The Virginia Marine Products Board reports that the United States Department of Commerce says, "Americans consume 16.5 pounds of fish and shellfish per person, with crabs and clams among the 'top ten' most popular seafood items."

About two dozen species of crabs are edible. Although in the U.S., we focus on just a few. The Chesapeake Bay blue crab is the most popular in this area. Stone crabs also are found in the Atlantic Ocean; however, the best-known fishing areas are the Florida Keys and Gulf Coast waters. One claw (sometimes both) is removed and the crab is thrown back in the water to regenerate a new claw, which takes about eighteen months. This is a sustainable way of harvesting claws. Claws can be harvested multiple times from the same crab without killing the crab. Their season is from October 15 through May 15, and if you have them any other time, they're probably coming from Chile.

*Dungeness crab cluster*

*Snow crab cluster*

The waters off Alaska's shores produce seven crab species that are caught commercially. They are the red crab, blue king crab, golden king crab, Tanner crab, snow crab, hair crab, and Dungeness crab. Additionally scarlet king crab, grooved Tanner crab, and Triangle Tanner crab inhabit the waters. They are caught between October and January, with restrictions on the number of days each crabber can bring in a catch. The crabs are frozen while at sea and then shipped around the world for year-round consumption.

You can see some of the trials and tribulations involved in harvesting these crabs in the TV series *Deadliest Catch* on the Discovery Channel. Just as with the blue crab, Alaska's crab stock has been depleted over the years. Unlike these boats that are at sea for weeks or months at a time, almost all of the catch from the Bay is caught on day boats that go out early in the morning and return when they've reached their limit and run out of hours.

One other crab you'll hear mentioned along the Eastern Shore and the East Coast is the horseshoe crab. Although it looks similar to crustaceans, it belongs to the Chelicerata family and therefore is more closely related to spiders and scorpions of the arthropod family. These crabs live primarily in shallow ocean waters on soft sandy or muddy bottoms. They will, however, come on shore for mating, where the female digs a hole in the sand and lays a few thousand eggs at a time. The eggs take about two weeks to hatch and most succumb to shore birds before they reach the hatching stage. They are commonly used as bait and in fertilizer, and in recent years there has been a decline in number of individuals, as a consequence of coastal habitat destruction and overharvesting along the east coast of North America. There is a moratorium against fishing for horseshoe crabs in New Jersey and Delaware.

Fossils of horseshoe crabs have been dated from 450 million years ago and have changed very little in the last 250 million years. A hard shell protects the entire body and the long rigid tail is used to flip itself over if it is turned upside down.

William Sargent, a consultant for the *NOVA* science series on PBS and former director of the National Aquarium, Baltimore, studied and wrote about the history and future of the horseshoe crab in *Crab Wars:*

*Horseshoe crab leaving the water (Courtesy University of Delaware, photograph by Lisa Tossey)*

*Horseshoe crab underside (Courtesy Calvert Marine Museum, Solomons, Maryland)*

*A Tale of Horseshoe Crabs, Bioterrorism, and Human Health.* He relates the story of the 300-million-year-old creature that is now involved in a battle between existence and the discovery that the crab's blood is a part of a reliable test for the deadly and ubiquitous gram-negative bacteria that cause meningitis, typhoid, E. coli, and other diseases.

Chesapeake Bay blue crab harvests have risen and fallen from the days when the crab was a trash food. Many factors are involved in its survival from how much snow melt there is in any given year to spring rains (when both are heavy, they lower the salinity of the upper Bay thus affecting how far north the crabs will venture). Run off pollution, algae blooms (they shade the underwater grasses, thus stunting the crabs' growth or killing them outright and adding to the pollution), overharvesting, and other factors all affect the welfare of the crabs.

The Bi-State Blue Crab Advisory Committee, the Chesapeake Bay Stock Assessment Committee, the Maryland Department of Natural Resources, the Virginia Marine Resources Committee, and many others gather to try to gauge the health of all things that in one way or another can affect the crabs.

Reportedly, crab processors say the restrictions are resulting in an annual sales loss of $13.5 million, with more than 450 processing jobs lost, and perhaps even, the closing of some small and maybe medium-sized processing plants. This is an industry that had fifty-three licensed distributors in 1995. With two-thirds of the processing plants located in Dorchester County (seven small, nine medium size, and all five large plants), on the Eastern Shore, and Dorchester already having the third highest rate of unemployment, this could be a serious economic loss to the county and the state. When you add the indirect income lost by can manufacturers, printers, and bushel barrel suppliers, the multiplier effect becomes enormous. Just as the 2010 oil spill in the Gulf of Mexico had a huge ripple effect on everyone from florists who provided bouquets to delivery drivers and others who supplied and benefited from the related industries. The economic downturn that has been plaguing the country and other economies has resulted in a slightly lower demand for the expensive product.

Another stab at easing the demand on crabbing was a program started in 2009 by the Maryland Department of Natural Resources and the Virginia Marine Resources Commission, which bought back 925 inactive crabbing commercial licenses. There are about 1,800 active crabbers in Maryland. Maryland paid $2,260 for each of 566 licenses while Virginia bought back 359 for varying amounts. These figures were as of January 2010. Although slightly successful, that still leaves a lot of people who seem to be holding on to their license to give or sell to a friend or relative or waiting to see if crabbing becomes more profitable again. Another 2,000 crabbers who didn't report any catch in 2008 had their permits frozen or restricted.

The funds, like the monies for the ghost trap program, came from federal disaster funds geared toward the recovery of the Bay. Each state was given $15 million for the various programs.

Of course, if the crab population were to dwindle to near extinction, whether by overharvesting or just a few years of bad Bay environment, that would have a detrimental effect, too.

Adding to that is the increase in the importation of crabmeat from overseas (Asia is a large competitor), where the crabs are about a third larger and the labor for picking is considerably less expensive. Notice the packaging when you purchase crabmeat and crab products. For example, the package may say Maryland-style crab cakes (or other products). Look for the oval USA American Blue Crab label to make sure you're buying American crabmeat.

The packing industry has had more than its share of notoriety in the past decade or two. Although there has been some success in a crab-picking machine, most of it still has to be done by humans. Most of the crab pickers at the packing plants on the Eastern Shore are Mexicans who are brought in under the H-2B visa program, which addresses employment needs during peak labor times for amusement parks, ski resorts, and other hospitality and entertainment functions. Maryland senator Barbara Mikulski, chairwoman of the Commerce, Justice, Science (CJS) Appropriations Subcommittee, has been a strong proponent of what she calls "Save Our Small and Seasonal Businesses Act," which regulates the number of immigrants who may enter each year.

The H-2B situation is a serious case of "he said, she said," and "they said." On the one hand, crab picking is tough work. The crabs that are picked are small or in new shells so there's not a lot of meat for the size or the meat can't be sold to the public. The pay is about $2.00 to $2.50 a pound of crabmeat, and you can get about eight pounds of meat from a bushel of crabs that weighs about fifty pounds. That's regardless of what size the crabs are. The pickers are paid a bonus for exceeding quotas and are provided with housing and some other benefits.

It may seem strange that foreigners are being allowed entry into the U.S. when the country is nearing 10 percent unemployment, yet it's work that most Americans just don't want to do. The department of labor institutes restrictions on the length of time a worker can be in the country in any one year (usually during the summer months when the crab harvest is at its peak). In order to qualify for the H-2B program, the employer has to file a recruiting report documenting that they tried to recruit local residents, provide information about each person who applied, and explain the lawful job-related reasons for not hiring each U.S. applicant.

However, not all employers treat their employees fairly. A study by the International Human Rights Law Clinic (IHRLC) decried the working and living conditions reported by forty immigrant workers during interviews the organization conducted. Some of the problems occur in Mexico, with local recruiters demanding a recruiting fee from female workers, despite laws that prohibit such payments. Generally, women are hired for picking because they have smaller fingers than men and can pick the meat from the small cavities better than men can. In the United States, rural and isolated living conditions are a problem, and in some cases, workers must rely on their employers for

*Jack Brooks, Cambridge, Maryland*

transportation into the nearest town. The report also mentioned the lack of modern conveniences afforded the workers, such as a stove and oven. Yet, people who work with these women on a regular basis respond that the women are from rural areas of Mexico and don't have experience using modern appliances.

Jack Brooks runs the J.M. Clayton Company, a crab-picking and seafood distribution company founded in 1890 by Capt. John Morgan Clayton, Brooks's great-grandfather. The company claims to be the "oldest working crab processing plant in the world." Jack's brothers, Bill, Joe, and Clay Brooks, the first of the fifth generation, operate the business.

*Crabmeat pickers*

*Crab cooking pots*

*Crab shells*

During the summer he will have five dozen female workers sitting at long tables picking crabs and packaging the meat into plastic containers that will be weighed, sealed, and refrigerated. More crabs are dumped on the tables as the piles dwindle. Jack says they'll "package 30,000 pounds of crabs and crabmeat each day during peak season." To provide customers with off-season crabmeat, Clayton's imports crabmeat from Indonesia and runs the picking and packing process all year, albeit at a slower pace in the winter. J. Clayton Brooks invented the first automatic crab-picking machine, which was patented in 1973. The Quik Pik takes meat from the shell at a speed of one hundred pounds of crabmeat an hour. However, manual labor is still needed to retrieve the most valuable lump or back fin meat. Jack says it would take twenty-five workers to equal the output of the Quik Pik. Just about every part of the crab is utilized, and stacks of shells are collected to be recycled into various agriculture uses.

To see what crab picking is like, watch the *Dirty Jobs* episode in which Baltimorean Mike Rowe visited J.M. Clayton Company. Jack says Mike was so impressed with this family operation that he returned with his family for a second episode taped at the plant.

# Bay Explorations

Bay heritage can be explored through the various museums that highlight specific aspects of the people who lived, worked, and enjoyed their lives here. Additionally, these museums offer sailing classes, decoy carving, festivals, and concerts. This list of museums is not all-inclusive, but it gives you an idea of their purpose. The first section is a directory of some of Maryland's nautically related museums.

Annapolis Maritime Museum, Annapolis, covers the three-hundred-year history of the port city, its dependence upon waterborne transportation, and its metamorphosis into the state capital and now, self-proclaimed, America's Sailing Capital. Two extremes of this city are ego alley, where owners of expensive yachts parade their financial goodies, and the little park across the street, where a statue of Alex Haley depicts the author sharing his life story to three children in the place where Kunta Kinte and other slaves were brought to be sold at a slave auction.

*Alex Haley statue, Annapolis, Maryland*

The Calvert Marine Museum, Solomons (www.calvertmarinemuseum.com), features the Drum Point Lighthouse, a touch tank, hunting for shark fossils from nearby Calvert Cliffs, and an exhibit of small sailing craft. The museum staff offers a seemingly endless calendar of concerts, discussions, and education programs.

At the Chesapeake Bay Maritime Museum, St. Michaels (www.cbmm.org), you can learn about skipjacks (the Maryland state boat) and oyster harvesting and climb to the top of the 1879 Hooper Strait Lighthouse. You can even boat to this museum and dock directly at their site.

Captain Salem Avery House, Shady Side (www.shadysidemuseum.org), is housed in Captain Avery's former home, dated from the late 1850s. The National Historic Landmark has some local history on display and features thorough aspects of the town's life over the years.

Havre de Grace Maritime Museum, Havre de Grace (www.hdgmaritimemuseum.org), where the Susquehanna River meets the Bay, traces the importance of canals to shipping and the Upper Bay.

Historic London Town and Garden, Edgewater (www.historiclondontown.org), is a huge continuing archaeological dig that is exploring the area's waterside history as part of the Anne Arundel County Lost Towns Project. You can see a garden as it would have looked when this was a thriving town.

*Hooper Straight Lighthouse, St. Michael's, Maryland*

*Salem Avery House Museum, Shady Side, Maryland*

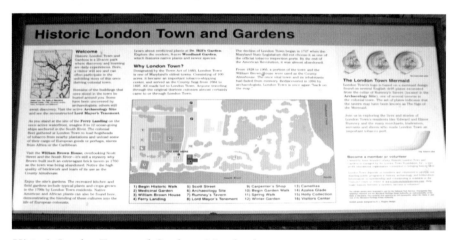

*Historic London Town and Gardens, Maryland*

Governor J. Millard Tawes Historical Museum and Ward Brothers Workshop, Crisfield (www.crisfieldheritagefoundation.org/misc.aspx?id=12), sits on the waterfront at Somers Cove Marina and explores the Bay's beginnings, seafood harvesting and processing, decoy carving and painting, and the contributions of the Native Americans.

Historic St. Mary's City, St. Mary's (www.stmaryscity.org), is an eight-hundred-acre living history museum on the grounds of the state's first capital in the seventeenth century. Despite the presence of the world-class St. Mary's College of Maryland, this is still rural country that hasn't seen the excessive overbuilding found in some other places along the Bay; the archeological finds have been outstanding.

Richardson Maritime Museum, Cambridge (www.richardsonmuseum.org), shows the heritage of wooden boat building with an outstanding collection of traditional Bay ship models. A nearby boatworks facility is teaching the next generation about the skills needed to craft a seaworthy boat.

The following museums are located in Virginia.

Hampton Roads Naval Museum, Norfolk (www.hmm.navy.mil), is housed on the second floor of the larger facility Nauticus (see below) in downtown Norfolk and focuses on the naval history of the area, including the role of the U.S. and Confederate navies during the Civil War.

Mariner's Museum, Newport News (www.marinersmuseum.org), has what is considered the "most extensive international maritime collection in America if not North America," with more than 75,000 volumes in the library and more than 350,000 naval-related photographs and images. One "WOW" exhibit includes the collection of miniature ships created by Winnifred and August F.

*Richardson Museum, Cambridge, Maryland*

Crabtree (former Hollywood set designer and model set builder). The USS *Monitor* Center is another must-see exhibit.

The Museum of Chincoteague Island (formerly the Oyster & Maritime Museum), Chincoteague, covers the maritime history of the island, the oystering and seafood business that is the island's major industry, and allows an up-close look at a first-order Fresnel lens that used to serve the nearby Assateague Light.

Nauticus, the National Maritime Center, Norfolk (www.nauticus.org), is a huge museum that has the benefit of being located next to the battleship USS *Wisconsin*. The center is home to a horseshoe crab cove, and one of the newest permanent exhibits is the Jamestown Exposition & Launching of the Steel Navy, which focuses on the period from 1880 to 1907.

*Nauticus Museum, Norfolk, Virginia*

Reedville Fishermen's Museum, Reedville (www.rfmuseum.org), is a tiny museum in a tiny town that covers the history of the Bay watermen and the menhaden fishing industry which was just huge in past days and is still a major factor in this area. The entire town could be considered a living museum, for the mile-long Main Street is a National Historic District that is flanked on both sides by water.

*Reedville Fishermen's Museum, Reedville, Virginia*

# Bon Voyage

Besides its recreational and commercial fishing value, the Bay is a major shipping lane into the Helen Delich Bentley Port of Baltimore, which serves two-thirds of the nation's population. According to *Logistics Management Magazine,* the port is "ranked number one in the U.S. for roll on/roll off cargo (farm and construction equipment), trucks, gypsum imports, and iron ore imports. It was ranked second for auto imports, sugar imports, woodpulp imports, and aluminum imports. And it was third in paper imports and wood imports."

It is one of the busiest and largest ports on the East Coast. The proximity of Interstates 95 and 70 means that shipping coming into this port makes it 125 miles closer for truckers and trains transporting goods to the Midwest and north and south. It's said that most Mercedes-Benz imports arrive through this port. The shipping channel, under the authorization of the River and Harbor Act of 1970, is maintained at fifty feet from Cape Henry to Fort McHenry. Rivers and streams north of Baltimore will freeze during particularly cold winters, so icebreakers are available to keep the shipping channel open through the most severe winters. (About 18 million cubic yards of silt dredged from the Bay's bottom have been used to restore Poplar Island, a disappearing spit of land just off Talbot County on the Eastern Shore.)

Baltimore has become a major cruise port, and the Maryland Port Administration reports that 210,549 passengers sailed on ninety-one cruises from the Cruise Maryland Terminal in 2010, exceeding the 2009 record of 167,235 passengers on eighty-one cruises. In a January 2011 news release, Gov. Martin O'Malley is quoted saying, "Baltimore continues to make waves as one of the top cruise ports in the U.S. The city's geographic location, entertainment options, and the cruise terminal's easy access right off Interstate 95 continue to attract passengers in record numbers. The Port of Baltimore's cruise business also pumps millions of dollars and generates hundreds of jobs for our state's economy."

Celebrity Mercury *cruise ship at Port of Baltimore.*

In 2011, 112 cruises were scheduled to depart from the port, with the *Carnival Pride* offering year-round trips to the Bahamas, Bermuda, and the Caribbean and Royal Caribbean's *Enchantment of the Seas* departing for the same area and New England and Canada. According to the release, Baltimore is now ranked sixth on the East Coast and fourteenth nationally for cruise passengers. Baltimore is also within a six-hour drive of 40 million people. The only physical constraint to even larger ships using the port and carrying more passengers is the 185-foot vertical clearance of the Francis Scott Key Bridge.

The twin spans Chesapeake Bay Bridge (officially the William Preston Lane, Jr. Memorial Bridge, named for the governor who initiated the original span's construction) crosses the Bay from Sandy Point, east of Annapolis (the state capital), to Kent Island and what is referred to as the Eastern Shore. Farther south, the Chesapeake Bay Bridge Tunnel connects Virginia Beach, Virginia, to Cape Charles on the southern tip of the Eastern Shore.

# Bernie Fowler

Clyde Bernard "Bernie" Fowler could easily be called the patron saint of the Patuxent River and the quality of life around the river and the Bay, for he has championed this cause since the 1970s. He says he could "talk forever about the river," which is the longest river totally within the state of Maryland.

He was born in Calvert (he pronounces it Culvert) County in 1924 and says, "The Patuxent River was a place I had a great deal of affection for, growing up there as a baby, child, and young man. I was a child of the Depression and I have very vivid memories about the kind of life we had then. With the abundance of aquatic life in the river at that time and during the Depression, we had plenty of resources and we didn't face the starvation of other areas.

"That river was so abundant with fish, crabs, oysters, and clams and we could always depend on something from that river. If it was frozen we could dig clams on the shore. There was good soil, so we had always had a garden growing. This all helped us through a very challenging time."

By the time the Depression was over, businesses and the economy were picking up. "It's unfair," he says, "the way we've treated the river. We've treated it as a waste receptacle, with over development, and coal-fired power plants spreading their poisons. Crabs are pretty hardy animals," says Bernie. "They can take a lot. They're scavengers and they'll eat anything that comes along, rotten eel—anything. The big problem is with the juveniles and the water quality has to be improved for them to survive. In the 1950s, we had trot line crabbers who would bring in as many as ten barrels of crab a day and they never kept any females. Me, my father, grandparents, aunts and uncles, never took females. Common sense told us that if we took the mamas, there wouldn't be any babies out there. The crab pots came into use and went bananas and as long as they were big enough, females were the biggest part of their market, frankly. In the past couple of years we've seen some things happen and I think it is going to pay big dividends unless it crumbles in the future."

Bernie gives a lot of credit to Harry Roe Hughes, the fifty-seventh governor of Maryland, from 1979 to 1987. Hughes was born in Easton, Maryland, on the Eastern Shore, and served in the U.S. Navy Air

Corps during World War II. During his two terms, Hughes signed the Chesapeake Bay Agreement, which started the fight against pollution and excessive wildlife depletion. When he could not run for a third term (constitutionally forbidden), he became a member of the Chesapeake Bay Trust from 1995 to 2003. Bernie says the organization filed suit against the Federal Environmental Protection Agency because the agency had not done what it was supposed to do under the guidance of the 1972 Clean Water Act.

Bernie was in politics, too, serving as a Calvert County commissioner (1970-1982) and Maryland state senator (1983-1994). Hughes, says Fowler, was able to convince the Waterman's Association of Virginia and other political and special interest bodies that the health of the Bay was important to all of us.

Despite what appears to be obvious benefits, Bernie acknowledges that there's a lot of opposition. "It's like the saying that everybody wants to go to heaven, but no one wants to die to get there. Everybody wants clean water until they find out it's going to cost them. A bill failed in Congress [in 2009], but I think Sen. Ben Cardin and Representative Cummings will be introducing it again this year that will codify the Clean Water Act of 1972.

"There are at least two lawsuits from Pennsylvania over the limitation of nutrients and toxic material and without that, there's no chance of cleaning the Bay."

Maryland senator Benjamin L. Cardin, chairman of the Environment and Public Works Water and Wildlife Subcommittee, presented a Chesapeake Bay Ecosystem Restoration Act of 2009 to Congress, which did not pass. According to Cardin's office, "It would have set a firm deadline of May 2020 for all restoration efforts to be in place." Cardin joined with Maryland governor Martin O'Malley, Congressman Elijah Cummings (MD-7), Congressman Chris Van Hollen (MD-8), and Virginia secretary of natural resources L. Preston Bryant (representing then Virginia governor Tim Kaine).

The original deadlines to clean up the Bay were set at 2000 and 2010 and obviously were not met.

At the time of this interview, Bernie was confident that Senator Cardin and Congressman Cummings would reintroduce the act in 2011.

One thing that was accomplished, says Bernie, "was the joint prohibition by Governors Kaine and O'Malley that restricted the catching of female crabs at certain times. The population of salable crabs increased 60 percent!"

If we don't clean the Bay now or start working on it, warns Bernie, we will reach a "point of no return or a tipping point and we won't be able to do it. We are racing against time and we're running out of it."

He's confident that good things are happening and says, "I think we have the people in place and the tools are in place. It's going to depend on the finances for getting the job done the way we should."

Bernie cites a study of Patuxent water quality that was deficient in its scope because "it only studied the upper river and decided that phosphorous was the villain in fresh water and stopped there. Farther downstream, in brackish water, nitrogen is the culprit. The economic situation hasn't helped, but it's no excuse for us to become complacent. We have to get our neighbors, Washington, D.C., Delaware, New York, Pennsylvania, and West Virginia, in line with what Virginia and Maryland are doing."

The history of the Bay clean-up is "one of a labor of love for a lot of us and they deserve a lot of credit for staying with it, for sometimes in the face of defeat, there's a cloud of discouragement and people become depressed and back off and our guys haven't done that and we know the prize is worth winning and we will continue running. We will snatch the Bay out of the jaws of defeat and bring it back to the 1950s.

"The National Association of Farm Bureaus has opposed this measure even though they don't know where the Bay is. People in the Plains states don't have that kind of interest, and the builders and developers have joined the farmers."

In Atlanta, Georgia, on January 10, 2011, American Farm Bureau (AFB) president, Bob Stallman, was quoted as saying the EPA's "pollution diet" was overreaching the agency's rights. Their argument, which to environmentalists does not hold much water, is that the states and not the EPA should regulate farming practices. Environmentalists argue that the states have not done a very good job, so someone else has to be in charge of the hen house. Stallman's comments seem to indicate the Chesapeake Bay Total Maximum Daily Load (TMDL) is a new regulation which goes back to the 1972 federal Clean Water Act.

In response, the Chesapeake Bay Foundation issued a statement stating, "This action by the Federation is not pro-farming action but anti-clean water." They went on to say that "clean water is important to farmers, their families, and their livestock."

One irony cited in the AFB's argument is the constant call for farm subsidies (or the federal government assisting in farm production) while saying the government shouldn't interfere with their farming practices.

In 1988 Bernie created the "wade-in" at Broomes Island to bring children, teachers, elected officials, environmental leaders, and the community together to focus on water quality. The event is held each year on the second Sunday of June. On this day Bernie marches into the Patuxent River and measures the depth of water through which he can see his white sneakers. He's pleased that in 2007, at the twenty-first wade-in, he could see through twenty-six inches of water, up from twenty-one inches the year before, which was up from eight inches in 1988. However, he recalls the "Sneaker Index" measuring more than fifty inches in the 1960s when he could walk into the river and be up to his shoulders and still see the bottom.

In 1998, the Bernie Fowler Laboratory was named in his honor. As part of the University of Maryland Chesapeake Biological Laboratory in Solomons, Maryland, the building has more than 25,000 square feet of research space, which the lab says is dedicated to the "study of environmental chemistry, organic and trace metal geochemistry, biogeochemistry, and microbial ecology."

Bernie quotes Sir Winston Churchill when he says, "No matter how you feel, no matter how many times you feel disappointed, it's okay—but you don't want to give up and for that reason, never give up, never, never, never give up—eight words but they're powerful. That bay could feed the world—it's coming."

# Crab-u-lous Time

Crabs, obviously, are a very serious business. On the other hand, we treat them with great humor. Perhaps it's because the season lasts only from May through November. We're giddy with delight during season, our interest drops off, and then peaks again after too much deprivation.

Jackie Leatherbury Douglass created a gorgeous stained-glass crab (said to be one of the largest stained-glass sculptures in the world) for Baltimore Washington International Thurgood Marshall Airport in 1984. The 10x7x5-foot sculpture *Callinectes douglassi* took about 5,500 hours to construct over a fourteen-month period and weighs about five hundred pounds. It's located in the main terminal, upper concourse (before security), housed in a protective case so onlookers and the crab can't be hurt or damaged.

Easton, Maryland, doesn't drop a crystal ball. Instead they have a Crab Drop, featuring a giant steamed red crab that's lowered as part of their First Night festivities, dropping at 9:00 P.M. (for families with children) and at midnight, on New Year's Eve.

*Crab sculpture at BWI airport. (Photograph by Marilyn DiMarco)*

Prior to the turn of this century, when everyone "knew" computers were going to crash because of Y2K, the millennium bug, or the Y2K bug, Richard and Suzanne Hood of nearby Royal Oak decided to avoid the domestic problems by spending some time in Bermuda. While visiting in St. George, they witnessed the traditional onion drop. It was on a six-foot pole, says Richard, and wasn't very exciting.

They thought Easton could do something similar, but with a crab and a little more pizzazz. Suzanne, an attorney and art major who was quasi-retired, decided she wanted to build the outer structure. The inner works are steel rods to give it shape, with chicken wire and papier-mâché. Richard, in the waste-equipment sales business, made the winding apparatus and the eyeballs and painted it with coats and coats of weather protection. By the time the idea was approved and the twenty-eight-foot pole and 8x4-foot crab (with claws sticking out overhead) were constructed, it was 2004 and about 450 children and family members showed up for the first drop. "Everyone wanted to touch it," says Richard, "and take pictures in front of it."

Richard stores the crab and apparatus in one of his garages, and he's turned over the crab operation to his son Lance to carry on the tradition. Richard admits the crab has become a little sad looking over the years and Lance is going to have to restore it. Richard is amazed, though, at how fast the word has traveled, for the Easton crab has

*Easton, Maryland, New Year's Eve crab drop. (Photograph by Richard Hood)*

been featured in national magazines and on Web sites as one of the New Year's Eve drops to see.

You can find candy dishes, rhinestone-encrusted jewelry, trivets, wall tiles, clocks, floor tiles, refrigerator magnets, clothing, wall hangings, sun catchers, mock street-crossing signs, what-not boxes, silly foam hats, sensible hats, coffee mugs, tote bags, bumper stickers, neon signs, embroidered tea towels, and cookie cutters in the shape of the crab or decorated with one. When Hollywood came east to film *State of Play,* with Russell Crowe, Ben Affleck, and Rachel McAdams, the crew brought a huge inflatable crab and placed it on top of the crab stalls at the Maine Avenue Fish Market. Mike Storm of Shoreline Seafood even has a crab tattooed on his leg.

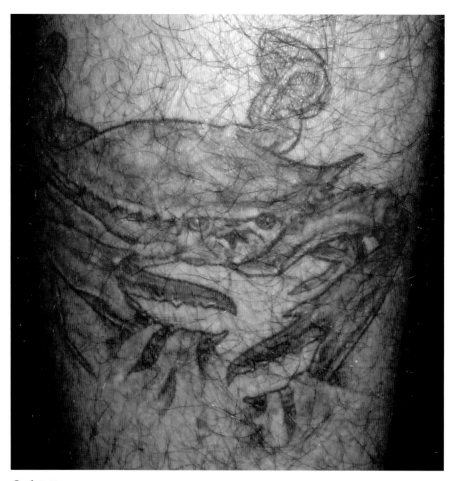

*Crab tattoo*

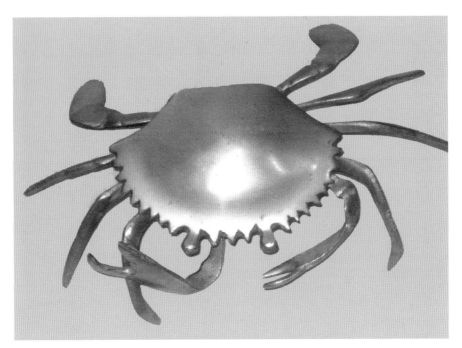

*Crab trinket box*

*The film* State of Play *was shot at Maine Avenue Wharfs, Washington, D.C. Crew members placed this inflatable crab above the fish market.*

*Crab wall clock*

*Crab feast coffee mug*

In 1981 David DeBoy was a stage actor and host of a children's show on WBAL-TV, the NBC affiliate. He also performed in TV and radio commercials and sang the occasional jingle. As DeBoy notes, his voice is not a highly commercial one. However, he did enjoy the earphones and singing to a musical track, so he figured he should write a song dealing with something local so local people would buy it. He combined Christmas and crabs, and with the help of Brent Hardesty (musical score), Jimmy Friedman (producer), and a bunch of friends, the song "Crabs for Christmas" was recorded and mixed in one day. Thirty years later, the song is a "Bawlamer" tradition. The song eventually made it to a program called *Crabs* that appeared on WMPT, the Maryland PBS station, and other performance venues, including a reunion of sorts at Geppi's Entertainment Museum.

The song relates the sad story of a Baltimore man sitting on Santa's lap in a mall north of Houston and saying all he wants is crabs. David probably could retire from the sales and royalties of that song, but he continues to entertain us, thank goodness. The chorus of the man's plea goes:

"Oh, I want crabs for Christmas.

Oh, only crabs will do.

*David DeBoy and the Hons, Karen Fitze (left) and Wendy Savelle (right), sing "Crabs for Christmas."*

Oh, ho, with crabs for Christmas,
My Christmas wish'll come true."

It seemed natural to invite David to contribute a recipe to this book. He says his wife, Joellen, is the real cook in the family, and they sent along a recipe for Maryland Crab Imperial Casserole and one for Crab Meltaways. You'll find both in the entrée section. David, however, does make crab-shaped cookies for Christmas, complete with red frosting. You'll find his recipe in the dessert section.

Chessie, a Bay snakelike creature (think Loch Ness Monster in Scotland or CHAMP in New York's Lake Champlain), was first reported in 1943. Various sightings have claimed it to be anywhere from twenty-five or forty feet long to seventy-five yards long. It may have a head that's about as big as a football. One guess is the monster is a misplaced manatee that strayed from the warm Florida waters it calls home. A few manatees have been captured in Bay waters and returned to the Sunshine State. As yet, the fishing and seafood markets have not considered Chessie and its possible relatives as commercially viable.

# Buying Crabs

You can buy hard-shell crabs already cooked or live (see more about soft-shell crabs below). Sometimes the cooked crabs cost a little more than the live ones. Male crabs (Jimmies) cost more than female crabs (sooks, mature females). Many think female crabs are sweeter or tastier while others believe male crabs are. Some people think female crabs shouldn't be caught so they can continue to live and produce more crabs. As mentioned earlier, Maryland and Virginia have had policies in place restricting crab catching in general and females in particular.

You can tell the males and females apart by the apron on the underside. The male apron resembles the Washington Monument. Mature female aprons favor the capitol dome, and immature female aprons are shaped like a triangle. Also, the male has blue-tipped claws while the female has red "nail polish" on the tips of her claws.

*Male and female crab underside*

Crabs, measured from tip to tip, come in various sizes (rough measurements—size names may vary depending on location, with a large in Louisiana being a medium around the Bay, etc.).

Colossal—6½ inches or larger
Jumbos—6 to 6½ inches
Large—5½ to 6 inches
Medium—5 to 5½ inches
Small—4½ to 5 inches

*Crab size comparison from (bottom to top) small, medium, large, and jumbo.*

They may also be called number ones, number twos, and number threes. Number one Jimmies will be the largest males, number two Jimmies will be smaller, and number threes will be a mixture of sizes and gender. Again, this will depend on where you are buying your crabs, geographically.

You may purchase a single crab or several bushels. The number of crabs in a bushel depends on the size, so a bushel of jumbos will have five to six dozen (60-72) crabs and a bushel of mediums will have seven to eight dozen (84-96). When feeding a crowd, figure on a half-dozen crabs per person if you're serving corn, salads, hot dogs, and other things. If it's only crab and beer, figure about a dozen each.

The individual crab will be more expensive than the per-crab cost of a dozen or bushel. Crabs purchased after July 4 will be much less expensive than before and even less expensive after Labor Day. Crabs are their heaviest and least expensive around October or toward the end of the season when "everyone" has tired of them.

When choosing crabs, you want ones with dirty or rusty bottoms. If the underside is bright and shiny white, then it has recently molted

*Rusty crab (top), clean crab (lower)*

and grown a larger shell which it has not filled. You're paying for a larger size shell but smaller size crab. You want "fat" or "heavy" crabs.

Donald Storm, owner of Shoreline Seafood in Gambrills, Maryland, and his brother Mike, say they gave the name of swamp dogs to the crabs that are even bigger than jumbos and can weigh up to 1½ pounds. (Ignore any information you read online reporting that crabs can grow to five or six pounds.) Someone asked Mike what the super large crabs are called, and Mike figured since they live in the swamps nothing bothers them so they grow and grow. Therefore, he named them swamp dogs. It's becoming a recognized size description although you're sure to find people who've never heard the term.

*Swamp dog crabs flanked by medium-sized crabs.*

# Catching Crabs

If the price of crabmeat or crabs doesn't please you and you decide you want to have a fun afternoon, ditch the electronics and communication gadgets and catch your own! You probably should time your crabbing with the incoming tide, so the "fun afternoon" should be taken figuratively. Find a pier or a bridge and hang a hand line or crab trap over the edge and just enjoy the outdoors (remember the sun block). Bring with you a pole net to retrieve the crabs off your hand line and a measuring stick of some sort to make sure the crabs are legal size. You also need a pail or pot with a cover or damp cloth over it to hold the crabs you catch.

The Maryland division of state documents lists the legal regulations regarding catching crabs in Maryland waters.

According to the Maryland Department of Natural Resources, if you are under the age of sixteen, you are not required to obtain a license for sport fishing or crabbing. However, if you purchase a recreational crab license, you are permitted to catch up to a bushel of male hard crabs and two dozen soft crabs or male peeler crabs or a combination of two dozen soft crabs and male peeler crabs under certain conditions a day. "A recreational crabber, of any age, may crab without a license all day, every day, from a dock, pier, bridge, boat, or shoreline if you're using a dip net or other type of handline. A property owner may set a maximum of two crab pots per privately owned pier at their property."

There's also a prohibition against catching crabs for recreational purposes on Wednesdays in the Bay or its tidal tributaries (except for crab pots from private piers) or on Thursdays if a state or federal holiday falls on a Wednesday or Thursday. You may not sell crabs caught for recreational purposes and you may not knowingly buy crabs caught for recreational purposes.

Furthermore, from April 1 through July 14, you may not possess a male hard crab that measures less than five inches across the shell from tip to tip of the spike, and after July 14, you may not possess a male hard crab that measures less than 5¼ inches across the shell from tip to tip of the spike. With a few exceptions, female hard crabs and female peelers may not be caught.

Yes, these rules can be very confusing. The easiest way to be sure is to check the current regulations when you buy your crabbing supplies and licenses.

However, the rules are different for Worcester County and the coastal bays of the Atlantic Ocean. For licensing purposes, the crab season is defined as April 1 to December 15. "A recreational Crabbing License is required of persons catching crabs for recreational purposes in the waters of the Chesapeake Bay and its tidal tributaries using (a) trotline not to exceed 1,200 feet in length (baited portion), (b) 11 to 30 collapsible traps or rings, or (c) up to 10 eel pots for catching the individual's own bait."

Licensing fees ranges from $2.00 to $10.00. If you are blind, a disabled war veteran, or were a POW, you are eligible for a complimentary lifetime license (except for the trout stamp). As regulations can change from one season to the next, contact the Fisheries Service for current regulations and if you have questions about sport fishing licensure at customerservice@dnr.state.md.us or by calling (410) 656-9526.

The following information is important: You "may not possess, transport, or pack a female crab from which the egg pouch or bunion has been removed, or an egg-bearing female crab known as the sponge crab."

You may catch crabs in the tidal waters of the Bay and its tributaries using a collapsible crab trap (one that is manually operated and is portable) that has a flat bottom no larger than 20x15 inches and with no more than four articulated sides. You may catch crabs with a dip net (a bag suspended from a frame with a handle), a handline, trotline (with restrictions), or a seine that is no more than fifty feet long and five feet wide. Other regulations cover commercial crab catches.

Suitable bait is as debatable as any other aspect of crab catching and eating. Some swear by rotten chicken necks or backs; others swear by eel. Some people go for fresh fish (mackerel, tuna, and squid) or a can of tuna fish with holes punched in the lid so the oil can escape. Others keep their bait recipe as secret as their crab-cooking recipes. In other words, read the rules and regulations if you plan to catch your own crabs. Otherwise, find a roadside truck, a crab shack, or other commercial crab house and buy your crabs there.

# Cooking Crabs

Just as there's more than one way to skin a cat, there's more than one way to cook crabs. The traditional Maryland way is to steam them. Not boil them, steam. In Louisiana and along the Gulf Coast, crabs are boiled. In Philadelphia, live crabs are cut in half and then thrown into a soup or stew pot with a bunch of other ingredients. Here, we steam them.

The Maryland Seafood Marketing and Aquaculture Development Program "does not recommend freezing steamed crabs because of the potential for bacterial growth." They warn against freezing because "the undigested food the crab has eaten and the wastes in the interior of the crab take a long time to freeze in home freezers and increase the risk of bacterial growth. We suggest the crab be cleaned first: remove the shell, legs, intestines, claws and fat. Only the meat-containing parts of the body and claws of the crab should be frozen. This frozen crab meat is best used for soups or casseroles."

*Crab steamers*

You can freeze any leftover crabmeat. It's best frozen in a prepared form, such as crab cakes, casseroles, and soups. You can vacuum pack crabmeat, but it will be best used in a dish that combines other ingredients with the crabmeat. Frozen crabmeat is best used within three months. There's so much liquid in the crab that freezing toughens it and then dries it out by the time you thaw it.

Soft-shell crabs can be frozen. Unless you're capturing the soft crabs yourself or are Johnny-on-the-spot during molt time, you're probably buying them frozen. If you should be fortunate enough to buy them live, cut off the face (behind the eyes) with a knife or scissors. Lift the top on each side and remove the gills. Remove the apron (on the underside). Wash the crab thoroughly and wrap securely in an air-tight bag or package. These, too, will store for up to three months.

### Basic Instructions per Dozen Crabs

1 crab steaming pot or extra-large steaming pot with insert and lid
1 pair extra-long tongs and/or heavy oven mitts or padded gloves

½ cup seafood seasoning
1 cup white vinegar or cider vinegar
3 cups beer or water (or mixture)
Live crabs
Additional seafood seasoning

Steaming is best done in a three-piece pot designed for the purpose. The liquid goes in the bottom pot, the crabs and seasonings go in the top pot, and the lid fits securely. So if you have a full pot, the little critters can't climb out of it. If you don't have a steamer, scrunch up some aluminum foil or put a trivet or other barrier in the bottom of the pot to keep the crabs out of the liquid. Then place the crabs and seasonings on top of the foil or trivet.

**Hint:** Some cooks drink the beer while it's cold, and just use the water and vinegar for the crabs.

Bring the liquid to a boil. Layer crabs in the pot. This may sound easier than it is. The crabs actually don't want to go in the pot. You may wish to remove small children and dogs' noses from the immediate vicinity and make sure you're not barefoot. (My first live crab experience involved some friends, an elderly dog that didn't move too well, and an old four-legged stove from some previous decade. A crab fell during the transfer period and scampered under the stove and the dog hightailed it to a dark corner of another room. Well, his tail was between his legs, and he hadn't moved that fast in years. I never saw the crab or the dog the rest of the day.)

Generously sprinkle each crab, or layer of crabs, with the seasoning as you put them in the pot. Cover and steam for 20 minutes or until the shells are bright orange/red. The actual time will depend on how many crabs you're cooking relative to the size of the pot you're using and the heat source (grill, stovetop, open fire, etc.). If your pot isn't

*Sabrina Flores coaxing a crab into a pot.*

large enough to hold all the crabs, then divide into batches and the spice/liquids to reflect the number of crabs you're cooking. As I frequently say, this is not rocket surgery.

DO NOT COOK DEAD CRABS. Do not store cooked crabs in the same basket that held the live ones.

To help avoid the little critters from pulling out or losing claws (fruitless self-preservation process on their part), refrigerate or dunk them in a bucket of ice water for a few minutes (until they're dormant) before cooking. Or you can kill them by poking the crab through the "head" with an ice pick or similarly sharp object. Just flip the crab (carefully) on its back and put the pick through the shell just below the top of the shell or behind the head/face.

Use tongs or at least tall cooking gloves to handle the live crabs and hold the crabs from the back near the swimmer fins (away from the claw end) to avoid being pinched.

Once crabs are cooked, eat immediately or wait until they cool a little. Refrigerate if you're planning to eat them cold. Do not mix cooked and live crabs in the preparation area. Do not use the same utensils for handling live and cooked crabs.

*Cooked crabs in bushel.*

The above measurements are not hard and fast. Some cooks throw in a bunch of salt—anywhere from a couple of tablespoons to a cup. As a general rule, I don't add salt. First, the critters come from a salty environment, so they're already at least a little salty. Second, if we're not sitting down to eat the crabs, but picking them for future use, you may not know how they'll be eventually used so you can determine how much salt you need when you use the meat in a recipe.

If you're cooking crabs for that most social of all eating events—a crab feast—cover a table with newspapers or torn-open brown paper grocery bags to help ease the clean-up afterwards. Have paper grocery bags nearby to collect the shells and other trash. Eating steamed crabs requires a slow pace and is not a race to the finish. It's a social activity with conversation, laughter, recalling old memories, and establishing new ones. It's best enjoyed by a crowd that may number three or thirty (remember, the crabs are cheaper per crab when you buy a large quantity).

While you're cooking the crabs, cook some fresh corn on the cob (either boil with or without husks or grill or prepare to your preference). Melt a stick of butter and put it in the well of a paint tray. Roll the husked ears down the tray ramp so it lands in the well and is thoroughly covered with butter. Have a pair of tongs and some corn holders nearby and await fresh corn heaven.

Serve the crabs with melted or clarified butter or vinegar or vinegar water for dipping the crabmeat. Or not.

Steaming variations: Use only water and spices or water, vinegar, and spices. Change the white or cider vinegar for red wine, balsamic, or other vinegar flavor you prefer.

Put all the mixture in the bottom and liberally sprinkle more crab seasoning over the crabs, about every dozen, and again on top.

Use Italian spices, Chinese five spice, or your favorite spices instead of the crab seasoning (you may want to cook them unseasoned and have the various spices available at the table). Depending on what you use, the amount of your spices, and the number of crabs, this can add a delightful flavor to the crabmeat, particularly if you'll be using the meat in a recipe instead of eating them.

Old Bay probably is the most popular or best-known crab seasoning around the Bay or maybe across the country. Developed in the 1940s by Gustav Brunn, a German immigrant, Old Bay is owned by McCormick & Company and has been for some time. Other crab seasonings that enjoy their own loyalty are J.O. Spice, Wye River, Blue

Crab Bay Co., and the more commercial Baltimore Spice (now Fuchs North America). Penzys Spices carries shrimp and crab boil spices. A few companies now produce New Bay, asking "How long has that can of Old Bay been sitting in your cabinet?" Chef Todd of Equinox restaurant mixes his own New Bay spices that can be purchased at Market Salamander in Middleburg, Virginia, or on their Web site.

You can taste these to determine which you prefer or you can make your own. These are the ingredients (pick and choose) that are most frequently used when the spice combinations are listed: coarse salt, crushed red pepper, dry mustard, ginger, black pepper, celery flakes, onion flakes, celery seeds, crushed bay leaves, laurel leaves, cinnamon, paprika, thyme, mace, black peppercorns, cardamom, mustard seed, cloves, nutmeg, and allspice. Put spices in a spice grinder or coffee grinder used only for spices and blend away.

Drinking water will not cut the spiciness of the crab seasonings. You need beer or milk to calm your throat. Also, be careful of rubbing

*Old Bay cans*

your eyes or nose while eating crabs that have been liberally coated with crab seasonings.

Don't limit your use of this spice mix to steamed crabs. Throw some on popcorn, potato chips, French fries, tater tots, fried chicken, or corn on the cob or in salads, eggs, and Bloody Marys.

Many people are allergic to the regular crab seasonings. If you are allergic to other crustaceans (lobsters, shrimp, crayfish, etc.), you may be allergic to crabs. A shellfish allergy (clams, mussels, oysters, and scallops) may or may not mean you're allergic to crustaceans.

When commercial crabbers, wholesalers, or retailers cook crabs, they steam them without any spices so you can do whatever you want with the crabs and crabmeat. When you're buying steamed crabs from a store, the roadside stand, or a crab feast, they most likely will have used crab seasoning, although you can request that they don't.

Elizabeth Fournier, from Cornerstone Funeral Services and Cremation in Boring, Oregon, will serve food to some of her grieving clients. She says she adores using "Chesapeake Bay Blue crabmeat in recipes ever since I ate them on a visit with cousins to Annapolis. This is a recipe I whip up when I know a Creole family will be coming to my country funeral home. I like to create dishes that suit the backgrounds of the visitors to my parlor." Remember, this is a West Coast recipe, and it calls for boiling. You can find Elizabeth's recipe for crab-dipping in the sauce section.

1 gallon water
1 onion, minced
2 celery stalks, minced
2 lemons, halved
1 jalapeño, halved
1 garlic head, halved
3 fresh thyme sprigs
2 bay leaves
2 tbsp. crab seasoning (Elizabeth calls these spices a crab boil)
½ dozen crabs
Salt to taste
Pepper to taste

In a large pot, combine the water, onions, celery, lemons, jalapeños, garlic head, thyme, bay leaves, and crab seasoning. Season the water with salt and pepper. Bring the liquid up to a boil. Add in the blue crabs and steam for about 15 minutes or until they've turned red.

# Picking Steamed Crabs

Watch a half-dozen people open a hard crab and you're likely to see a half-dozen ways to complete the procedure. One person starts by removing the apron and then the back shell, cleaning out the innards, splitting the crab in half between the chambers, and then removing the legs, claws, and swimmer fins. Or you can remove the apron and shell and then the legs, etc. Or you can do the following:

First, break off back fin legs and remove any clumps of meat attached to them. Then break legs at joint and remove meat, squeezing the legs will help. Break off claws, crack at joint, and pull apart; tap the edge of a knife with a mallet to get at meat but don't smash the claw or leg. Turn crab on its back and pull back the apron. Insert your thumb at that point and pull off shell. Remove gills or devil fingers and dispose. Remove the stuff in the middle between the two sides of chambers of the crab body. Split the two sides apart gently. Take a knife and cut each half in half between the two sides of the cartilage. This will reveal the inside of the chambers and make it easy to remove the meat.

Of course, you can eat the crab right then and there. Some people use melted butter, vinegar, mayonnaise, or more seasoning as a condiment. Or set the meat aside and try dipping the meat into one of the recipes in the sauce section.

If you open a crab, whether you cooked it or bought it commercially steamed, and it's full of water, it's not from improper steaming (this does not include if you boiled them). What's happened is you've come across a crab in a new shell and the crab inflates itself with water to fill the difference in size between it and the shell. Remember, ask for "fat" crabs or look for ones with rusty or dirty abdomens.

You can eat the yellow stuff or mustard of crabs, if you're sure you know the purity of the water from which the crabs were caught. However, this is what purifies the crab's system of toxins and other nasty elements. Even if you know the water quality, there are warnings against eating the mustard if you're pregnant, of child-bearing age, or if you're under the age of five. If your crabs come from advisory areas, you probably shouldn't eat this part of the crab or drink the water in which the crabs are cooked. If you know the water is safe and you're the almost-daring type, you can add the mustard to your crab recipes as it imparts a slightly tangy flavor.

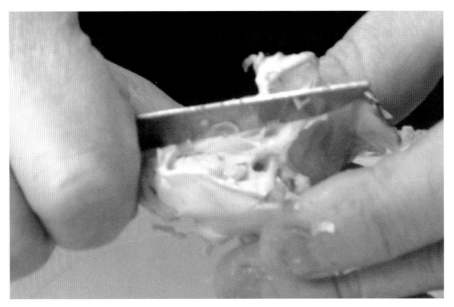

*Cutting half crab chamber to access lump crabmeat.*

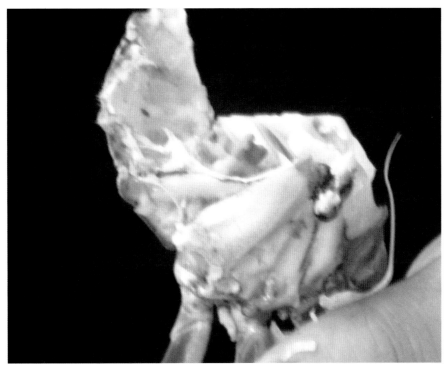

*Half crab chamber with lump meat*

# Soft-Shell Crabs

Soft-shell crabs occur when the crab sheds or molts its outer shell, which the crab has outgrown. The crab must be removed from the water within a few hours of shedding to prevent the tender skin from hardening into another larger shell. Soft-shell crabs are sold whole. They're in season from May to mid-September and vary in size from four to six inches. The first molt is said to coincide with the first full moon in May.

To clean a soft-shell crab, cut away the eyes and mouth with a sharp knife or scissors. Fold back one side of the top shell to expose the devil fingers (gills) and remove from both sides of the crab. Any frozen soft-shell crabs, available all year, you buy will have been cleaned before being frozen.

Prior to molting, crabs fatten up and are intensely flavorful. Soft-shell crabs should be eaten within four days of the molt before the shell starts forming. Molting generally lasts from early May to July.

Crabmeat is available year-round. After it's picked, it's pasteurized in hermetically sealed cans with no preservatives or additives (except for machine-picked claw meat). It will last up to six months when properly stored unopened in the coldest part of your refrigerator. This would mean, usually, the bottom shelf, toward the back, which is fine as long as you remember that it's there.

Some people will mutter that pasteurized crabmeat just doesn't have the same taste or texture as fresh crabmeat. According to Jack Brooks of J.M. Clayton Co., years of study have shown that about 51 percent of the population can truly tell the difference between the two. Of course, there are some people who can't tell the difference between crabmeat and imitation crabmeat (surimi) when thrown into a dish with other ingredients.

Fresh crabmeat, sold in plastic containers, should also be stored in the coldest part of your refrigerator and used within a week to ten days.

# Glossary

Here are crab-related terms you'll find throughout the book.

**Apron:** the abdominal (bottom) covering of the crab. In the immature female, it is triangular in shape. In the mature female, it looks like a capitol dome. In the male, it looks like the Washington Monument. The eggs are carried under the female's apron. In some areas, the apron is called a tail.

**Back fin:** the swimmer or bottom fin. The crabmeat connected to it is very muscular and it's where back fin crabmeat is taken. These pieces are smaller than lump or jumbo lump meat.

**Carapace:** the shell or hard covering of the crab.

**Chicken necks:** a popular bait used by recreational crabbers, with the thought that the older and stinkier, the better. The necks are tied to the bottom of a collapsible trap, a trotline, or hand line crabbing off the edge of a pier. These crabbers may be called "chicken neckers," and it is not a term of endearment, although sometimes it may be said in jest. Some people swear by them, others will use turkey necks, eel, or almost any type of fish. Crabs will eat or investigate just about anything.

**Crab pot or crabpot:** a large square trap made of coated wire similar to chicken wire. The pot has two chambers, the bottom or downstairs and the top or holding area. The crab enters the bottom part through a funnel, which makes it difficult to escape. The crabs swim upwards through another funnel to the top chamber, with the funnel again making it difficult to exit. These are used by commercial crabbers and are baited and left unattended for twenty-four hours or longer.

**Crab trap:** this is a smaller square trap, about one foot square, with some as small as six inches high, that is also made of coated wire. The bait is tied to the middle bottom of the trap and the sides are hinged so when it's lowered, the sides fall onto the bottom of the Bay or river. When the trap is pulled up, the doors are closed, capturing the crab inside. Recreational crabbers who enjoy spending several hours or a day catching the evening's dinner use these traps. The biggest problem with crab traps is that you have to raise the trap about every ten minutes to see if any crabs have entered. Otherwise, they'll become disinterested or finish eating the bait and leave.

**Devil fingers** (or sometimes Dead Man's fingers): the spongy-looking gills (not lungs) under the shell that are removed before cooking or eating. They aren't poisonous, but they taste funny and have a bad texture.

**Dip net:** this is a six-foot or eight-foot pole with a sixteen-inch wire loop at the end from which a net bag is attached. When you spot a crab swimming near where you are, reach into the water with the net and grab that little critter. Alternatively, if you're using a line with bait at the end of it, as you pull up the line (see hand line) and see that lovely crab at the end, swoop down with the dip net and capture the crab before it lets go of the bait.

**Ecdysis:** the shedding of the old shell when it is outgrown and the crab is molting. The new shell will become hard again in about two to three days. You should be able to see a pink line along the inner border of the back fin or paddle about a week before the molt.

**Fat crabs:** crabs that have grown into their new shell; the bottoms will be "rusty" or "dirty" looking compared to crabs in new shells that are very white on the bottom. When compared in weight, the fat crab with the rusty bottom will be much heavier than the one with a clean, white bottom.

**Handline:** a piece of string (solid string, not thread) with bait tied to the end that is drowned in hopes of attracting a crab to the bait. The line is raised every few minutes or whenever you feel a gentle tug and a dip net is used to capture the crab that your bait has attracted.

**Jimmy:** male crabs.

**Length:** the measurement between the two outermost edges that are farthest apart. One might logically think this is the width, but go with length.

**Mustard:** the yellow "stuff" inside the crab that filters impurities from the crab's system. If you don't know the waters of your crabs' origins, you probably shouldn't eat it. If you're confident your crabs came from non-contaminated waters, then feel free to eat it or add it to salad, soup, or other mixtures.

**Paper shells:** another term for a soft-shelled crab after it has molted, as the new shell is starting to harden.

**Peeler:** a crab that is shedding the shell it has outgrown.

**Sally:** a sexually immature female crab. The apron (on the underside) is triangular in shape.

**Soft-shell crab:** a crab that has shed its hard shell before the new shell starts to harden, about twelve hours after the molt. You eat the entire crab minus the devil fingers and head area. The soft-shell crab season starts with the first full moon of May and continues through September.

**Sook:** a sexually mature female crab.

**Sponger:** a female crab that is carrying eggs. (It's illegal to catch or possess spongers.)

**Swamp dog:** an extremely large crab, named by Donald and Mike Storm of Shoreline Seafood in Gambrills, Maryland, to describe super large crabs that grow in the swamps and marsh ponds of southern states. Also called a marsh crab, for obvious reasons. In other areas, it might be called a mumbo as a combination of massive and jumbo.

**Trotline:** a line, anchored at both ends, that's baited every two to six feet with shorter pieces of line attached to the main line, used by commercial crabbers. The line is played out as the boat motors along and then reeled in over a roller that pulls any attached crabs to the surface where they can be netted.

**Whale:** an especially large crab.

## Hard Crab Market Terms

**Bushel:** a basket of hard-shell crabs, weighing about fifty pounds, and containing anywhere from three to eight dozen crabs, depending on their size.

**Dozen:** crabs are generally sold in the half-dozen, dozen, or multiple dozen quantities until you reach the half-bushel or bushel order.

**Mixed:** a blend of crab sizes that may be mediums and large or some other combination. Whether they are "fat" is a more important criterion than absolute size.

## Types of Crabmeat

**Back fin:** large white pieces of crabmeat from the back or swimmer fin cavity. The pieces are smaller than lump or jumbo lump.

**Blended back fin:** a blend of back fin and special crabmeat, about a one to three (back fin to special) ratio.

**Claw:** darker than back fin or lump and sweeter and richer in taste. It's great for soups, casseroles, and other dishes that are more concerned about the taste than the appearance. It is the least expensive type of crabmeat.

**Cocktail Claw:** this is the meat from the large claw when the pincer is pulled out. It's sold in cans or as an appetizer tray a grocery store, fish store, or restaurant has prepared. The claw part remains intact to be used as a "handle" for dipping into cocktail or tarter sauce. You may find it listed as crab fingers.

**Dressed or pan ready:** soft-shell crabs that are ready to cook and eat; the eyes, mouth, gills, and apron have been removed. They're sold live only during soft-shell season (May to September); otherwise, they're available frozen.

**Lump or jumbo lump:** the prize, in matters of appearance, for white meat lumps taken from the chamber to which the swimming or back fin is attached. This is the most expensive crabmeat you can buy and generally has no shell or cartilage.

**Mixed:** a combination of lump and special crabmeat.

**Pasteurized:** crabmeat that has been picked from the cooked crab and then heated in hermetically sealed containers so all the harmful organisms are destroyed. The refrigerator shelf life is longer for pasteurized crabmeat than for non-pasteurized crabmeat. In taste tests, about 51 percent of the people surveyed could accurately identify which crabmeat was pasteurized and which wasn't.

**Special** (also called regular, deluxe, flake, or white): smaller pieces or bits of white crabmeat from the chambers of the walking legs. It rarely includes back fin and does include more shell and cartilage pieces because the chambers are tinier than the back fin chambers and it's more difficult to pick. Special is the least expensive white crabmeat.

# Food Tips

**Herbs and Spices**

In the battle between fresh and dried herbs and spices, your decision depends on convenience, taste buds, and cost. International commerce being what it is these days, you can generally find fresh herbs at your neighborhood grocery store. If not, you can find them packaged at the store or online. If you're talented enough, you can grow most of your own. The Chesapeake Bay area isn't like living in Southern California where windowsills regularly have little garden pots on them, but it can be close.

If you aren't the green-thumb type, buy dried herbs. Always keep them in an air-tight container and away from the light and heat. Yes, sounds strange because you have heat and humidity in the kitchen, but try. Make sure they're fresh by crushing a leaf or two to see if the aroma is still strong.

A chef acquaintance of mine went through a basil stage where he put basil in everything, including eggs, sandwiches, salads, and entrées. As I spent a lot of time enjoying his company and dining with him, I lost my zeal for that spice. I still have trouble eating pesto and anything else that even looks like basil was near it. If you're like that about a particular herb or spice, eliminate it from the recipe or replace it. The recipes featured in this book are suggestions of interesting ways to eat crabmeat at whatever time of day you prefer.

Here are some basic herbs and spices that you're likely to come across in this book:

Bay leaves—Very strong; use whole while cooking and remove before serving.

Celery seed—Strong taste, similar to the vegetable. Use it in moderation, sometimes ground in a spice mill with other spices.

Chives—Similar to onions, only sweeter and milder. It's usually sliced very thin and often added as a garnish.

Dill—Both the seeds and the leaves are tasty, particularly when used in soups, dressings, and with fish.

Paprika—A bright red pepper, it's a main ingredient in crab seasoning and, therefore, in almost every dish with crabmeat.

Parsley—Curly leafed or Italian, the preference is for fresh, but dried will do. Incorporate it in a recipe or use it as a garnish.

### Flour, et. al

Some people are allergic to ingestibles. If you're allergic to shrimp, you may also be allergic to crabs, lobsters, and crayfish. I'm allergic or sensitive to barley (beer and whiskey), cocoa, dairy, scallops, wheat, and walnuts. So far, I haven't had to go to the hospital after eating any of these foods, but I know they can make me uncomfortable, particularly the walnuts and scallops.

If you know, or determine, that you're sensitive to a specific ingredient, then eliminate it or replace it. For the most part, the primary allergen in this book (other than crabmeat) will be wheat or gluten. I've suggested a few wheat-type products that you can try substituting for processed flour. Some recipes, crab cakes for example, call for coating the cakes with flour or seasoned flour. You'll be fine eliminating that step.

Depending on where you live and shop and how plain or fancy you like your food, you have several options when it comes to the flour you use in these recipes. Unless otherwise specified, feel free to experiment when you're cooking; pay attention when you're baking.

*Back fin crab cakes sign*

## Flour Types and Other Thickeners

**All-purpose:** a blend of hard and soft wheat that may be bleached or unbleached.

**Buckwheat:** adds a nutty flavor and is good for people who are gluten intolerant.

**Cake:** good for fine-textured cakes, some quick breads, muffins, and cookies. If you don't have cake flour around, you can use all-purpose but deduct two tablespoons for each cup called for in the recipe.

**Chickpea or gram** (not graham): made from ground chickpeas and used frequently in cuisines from India, Pakistan, and Bangladesh.

**Instant blending:** Wondra (from Gold Medal) or Shake & Blend are great for sauces and gravies (for newbies) because it rarely lumps.

**Pastry:** also good for baking and is somewhere between all-purpose and cake flours. Sometimes referred to as cookie flour. You can somewhat simulate pastry flour by mixing two parts all-purpose to one part cake flour.

**Potato** (or potato starch): this is another gluten-free starch, which is considered kosher for Passover—it isn't a real issue in a cookbook that features non-kosher crabmeat.

**Rice:** is made from white or brown rice rather than wheat.

**Self-rising:** has salt and leavening mixed into the flour and is used for biscuits and some breads. It eliminates the need to measure and add salt and baking powder. You can mix your own by adding 1½ teaspoons of baking powder and ½ teaspoon of salt to each cup of all-purpose flour.

**Semolina:** used primarily for pastas.

**Whole wheat:** higher in fiber and lower in gluten than white flours.

Another area of choice and some confusion deals with salt. As many people are on salt-restricted diets, the amount you use can be important. Here is a little low-down on sodium chloride or NaCl. The main differences that you can see between salts are the size and texture of the salt granules. Table salt dissolves quickly (good for baking) while sea salt and kosher salt are a little briny and definitely crunchy, particularly when added at the end of your preparation. Some people feel the size of kosher salt makes it easier to grab a "pinch" of salt when cooking. Also, because table salt is finer than kosher or sea salt, there will be more table salt in any specific measure (e.g., teaspoon) than the other salts.

Table salt comes from underground mines and has a little calcium silicate to prevent clumping. People "of a certain age" will remember putting rice in the salt shaker to prevent clumping, particularly in areas with high humidity, which could include a kitchen where a lot of cooking is done.

Sea salt comes from seawater that's been evaporated so it contains the chemicals that were in the evaporated water. These minerals alter the taste and the color, although both attributes are lost in the cooking process.

Kosher salt, used in the koshering process, is obtained from underground mines or from evaporated water, has no preservatives, and with it's large crystal size is terrific for absorbing moisture when you're preserving food.

While most measurements in this book are consistent, there may be some instances when a recipe calls for two ounces instead of ¼ cup or vice versa, so here's a quiet refresher chart in case you come across a measurement you need converted.

### Measurements

| | | | | | |
|---|---|---|---|---|---|
| 1 cup | = 8 fluid ounces | = | 16 tablespoons | = | 48 teaspoons |
| ¾ cup | = 6 fluid ounces | = | 12 tablespoons | = | 36 teaspoons |
| ⅔ cup | = 5⅓ fluid ounces | = | 10⅔ tablespoons | = | 32 teaspoons |
| ½ cup | = 4 fluid ounces | = | 8 tablespoons | = | 24 teaspoons |
| ⅓ cup | = 2⅔ fluid ounces | = | 5½ tablespoons | = | 16 teaspoons |
| ¼ cup | = 2 fluid ounces | = | 4 tablespoons | = | 12 teaspoons |
| ⅛ cup | = 1 fluid ounce | = | 2 tablespoons | = | 6 teaspoons |
| | | | 1 tablespoon | = | 3 teaspoons |

Some recipes will offer suggestions for alternative ingredients to help please your taste buds. Others won't, but that doesn't mean you can't substitute one item for another. For breading in a crab cake, you can choose crumbled, stale bread; crackers; canned breadcrumbs; Italian breadcrumbs; Japanese panko breadcrumbs, or whatever kind of crumbs you prefer or have on hand.

Mark Ainsworth, professor at the Culinary Institute of America in Hyde Park, New York, says you rarely see cheese and fish together in a dish "because the cheese tends to overwhelm the mild flavor. There are exceptions, including Lobster Thermidor, which is finished with Parmesan cheese, and several Greek shrimp dishes that contain feta. With that said," he continues, "the number one fish and cheese combo in the world is McDonald's Filet of Fish, and you could argue that it is not really cheese. The reason is that the delicate flavors of fish collide with the strong flavors of cheese. They are just not complementary and tend to work against each other."

What's true of the fish is true of the crabmeat. However, if you want to try adding a little cheese flavor, try incorporating Parmesan cheese to batter when frying, or if you want cheese in your dish, just start experimenting to see what combination you like. There is a recipe for macaroni and cheese with crabmeat in the entrée section.

Several recipes call for lemons or lemon juice and if you like the citrus taste and want to experiment, you might want to try Meyer lemons that are sweeter than other lemons and have a slight taste of honey and thyme. The potential problem on the East Coast and around the Bay is the Meyer lemon season is roughly November through March so you're either going to use out-of-state crabs or packaged crabmeat. You also can try pink lemons just to be a little different. Limes will give you a slightly different taste.

A vegetable receiving a lot of culinary attention is the icicle radish, which is related to the turnip and horseradish family and has a crisp texture and a peppery-hot flavor. The icicle is a spring harvested radish, so it times well with crab season.

Unlike baking, where certain foods react with other products to produce a result (yeast in bread, for example), most cooking is not rocket surgery. Most recipes can be doubled or tripled without any problems.

Replacing ingredients and correcting mistakes can be a little tricky, but not necessarily impossible. Remember, chocolate chip cookies, French dip sandwiches, Popsicles, brownies, and fried ravioli are all said to have been created by accident or by making good of something that wasn't right.

## If you

- Have a tomato-based food that's too acidic, add baking soda, a teaspoon at a time. Or try sugar, again, a little at a time.
- Are out of tomato juice or sauce, mix equal parts of ketchup and water or to the consistency you want.
- Have lumpy gravy or sauce, put in the blender, food processor, or use a hand-held blender. You can also try straining out the lumps.
- Have too-thin soup, add flour a tablespoon at a time mixed with an equal amount of cold water so it doesn't lump. Or use Wondra that doesn't lump. This also works if your sauce or gravy is too thin. Cornstarch, tapioca, and arrowroot are other acceptable thickeners

that should also be mixed into a slurry. Cornstarch is best for dairy foods. Arrowroot thickens very quickly, so make sure you know how thick your soup/sauce is before you add more. Arrowroot is not a good choice for dairy foods because it becomes slimy when mixed with them.

• Have salty food, add sugar and cider vinegar, a teaspoon at a time. If your soup is too salty, add a potato that's been cut into large pieces and simmer for about 15 minutes and then remove the potato.

• Have greasy food, chill so the fat comes to the top and remove with a slotted spoon.

*Crab bushels*

## Cooking Terms

Although most cooking terms are defined within a recipe, a few might have escaped this redundancy. This list should help.

**Au gratin:** dish topped with cheese (and sometimes crumbs) and browned in the oven or broiler.

**Bisque:** a thick cream-based soup.

**Crudités:** raw vegetables, including broccoli, carrots, cauliflower, celery, and mushrooms that are served as hors d'oeuvres with a dip (crab dip, in this case).

**Dredge:** coating soft crabs, fish filets, and other foodstuffs lightly with flour, corn meal, or other product when frying. Sometimes you coat with the flour, dip in well-beaten eggs, and then dredge again. When seasoning the coating, the seasoning(s) should be in the flour mixture.

**Fold:** incorporating a delicate item (usually beaten egg whites or whipped cream, but in this book mostly the back fin or lump crabmeat) into another substance without breaking the air bubbles (in the whites and cream) or the lumps of meat. Put the crabmeat on top of the rest of the mixture and using a rubber or silicon spatula, bring part of the bottom mixture up over the crabmeat. Turn the bowl a third and repeat and continue until the ingredients are blended.

**Julienne:** cutting vegetables or other ingredients into thin strips, about ¼ inch and anywhere from 2 inches long and up, depending on how they'll be used.

**Sauté:** cooking or browning food in butter, olive oil, or other fat or shortening.

**Simmer:** cooking on a low heat, short of the boiling point, so the surface is only occasionally broken by a rising bubble.

# Blue Crab Festivals

As you should realize by now, half the fun of eating crabs is socializing. When you don't want to bother with buying, catching, cooking, and all the other aspects of having people over to share your crabs, then head to a crab festival or feast.

The following is a list of a few annual crab festivals and contact information. Before you take off for a day of enjoyment, call or check their Web site just in case any details have changed. Check the Maryland and Virginia tourism calendars or do an Internet search to find the dozens of crab feasts held almost every summer weekend, sponsored by civic associations, religious groups, and other gatherings to raise funds or just have a good time.

Memorial Day weekend
**Annual Soft-Shell Spring Fair**
Local seafood (soft-shell and crab cake sandwiches), arts and crafts,

*Crab feast (Courtesy Shoreline Seafood, Gambrills, Maryland)*

Waterman's Hall of Fame awards, entertainment, and more. Sponsored by the Crisfield Lions & Lioness clubs.
   City Dock
   Crisfield, Maryland
   (410) 968-2500
   www.crisfieldchamber.com/events.htm

Mid-June (Sunday)
**St. Mary's Crab Festival**
Seafood, automobile show, demonstrations, arts and crafts, farm animals, music and dancing, and more. Sponsored by the Leonardtown Lions Club
   St. Mary's Fairgrounds
   Leonardtown, Maryland
   (301) 475-4200 ext 1404
   www.stmaryscrabfestival.com

Mid-July
**Taste of Cambridge Crab Cook-off**
Competition for the best crab dishes by area restaurant chefs.
   Downtown Cambridge, Maryland
   (410) 228-0020
   www.cambridgemainstreet.com

Mid-August
**Chesapeake Crab and Beer Festival**
Taste crab dishes and beer.
   National Harbor, Maryland
   (301) 203-4170
   www.mdcrabfest.com

Labor Day weekend
**National Hard Crab Derby & Fair**
A derby, parades, beauty pageants, boat races, arts and craft exhibits, crab-picking contests, and more. A cooking contest is open to amateur chefs/cooks only.
   Crisfield Area Chamber of Commerce
   PO Box 292
   Crisfield, Maryland 21817
   (410) 968-2500 or (800) 782-3913
   www.crisfieldchamber.com/events.htm

Mid-September
**Maryland Seafood Festival**
Best of Maryland seafood, entertainment for the family, and arts and crafts.
  Sandy Point State Park, Maryland
  (410) 268-1437
  www.mdseafoodfestival.com

First weekend in October
**West Point Crab Carnival**
Games, arts and crafts, antiques and merchant booths, parade, and fun run/walk and tot run.
  West Point/Tri-Rivers Chamber of Commerce
  621 Main Street
  West Point, Virginia 23181
  (804) 843-4620
  E-mail: wptrcc@oasisonline.com
  www.crabcarnival.com

The Annapolis Rotary Club has been holding a humongous crab feast every August since 1946 and it keeps getting bigger and better, regardless of the weather (most tables are protected from the weather).

*Overhead view of crab feast. (Courtesy Shoreline Seafood, Gambrills, Maryland)*

They claim it's the biggest in the world and I'm not about to argue with them.

The 2010 crab feast resulted in $70,000 being donated to local charities and non-profit groups. Tickets for the 2011 feast were $60 for adults and $20 for children (3-12 years old). You can also buy VIP tickets that provide you with seating in a dining tent and a wait staff to bring the trays of crabs to you. The Navy-Marine Corps Memorial Stadium in Annapolis is the venue for the feast.

Mike Storm, of Shoreline Seafood, says they've been providing the crabs for the feast for at least the last decade. According to the Rotary Club, at the 2010 event, the 2,500 people in attendance consumed 478 bushels of crabs, 3,400 ears of corn, 130 gallons of crab soup, 1,800 hot dogs, and 150 pounds of beef barbecue (the dogs and barbecue came from Adam's Ribs).

For more information, contact the Rotary Club at (410) 353-4722 or visit their Web site at www.annapolisrotary.org/crabfeast.

*Mike Storm, Shoreline Seafood, Gambrills, Maryland*

# Part II: Recipes

# Breakfast

I'm a firm believer that lunch, dinner, and picnics should share crabs with breakfast. Some people want breakfast as their first meal, no matter what time it is. My friend, the late Saul Fruchthendler was one of those folks. He would be in crab heaven with these recipes.

With these featured dishes, you can eat breakfast all day.

*Crab dish towel*

# Crab and Corn Muffins

I'll admit, the first time I heard someone talk about crab muffins, I thought they meant crabmeat incorporated into the muffin mix. Instead, the recipes I found and tasted were English muffins, split in half with a crab imperial mixture (or similar) on top of the muffin and broiled.

A three-step crab-topped English muffin recipe follows.

This is a crab muffin, with crabmeat in the batter. You can use whatever kind of crabmeat you want. Lump meat will be a couple of surprises as you eat the muffin; special will impart a slightly stronger crab taste. I generally include the crab seasoning, which is not enough to make the muffins hot or spicy, but just enough to give it that familiar bite. This recipe calls for self-rising flour so leavening ingredients are not needed.

Although these recipes are in the breakfast section, you could cut wedges or one-inch pieces of French or Italian bread (in the three-step recipe) and serve as an appetizer.

Yields 12 muffins

½ cup polenta
½ cup liquid from any juice from the corn can and milk
Spray vegetable oil
4 strips of bacon, cooked and chopped (save the grease to use for sautéing the onions)
4 scallions or green onions, chopped fine, both the green stalk and the white bulb
1½ cups self-rising flour
1 tbsp. sugar
¼ tsp. crab seasoning (optional)
Pinch of salt
8 oz. fresh corn off the cob or a can with the liquid drained and saved
3 oz. butter, melted
2 eggs, lightly beaten
¼ cup Cheddar cheese, grated, or other hard, mild cheese of your preference
1 tsp. fresh dill
6 oz. crabmeat

Add the polenta and the liquid to a bowl and let it sit for about 10 to 15 minutes.

Spray a 12-hole muffin pan with vegetable spray (butter flavor tastes particularly good) and set aside.

Sauté the onions in the bacon grease or other fat/oil of your choice.

Add the flour, sugar, crab seasoning, salt, corn, butter, eggs, bacon, onions, cheese, and dill to the polenta mixture. Gently incorporate the ingredients and then add the crabmeat.

Spoon the batter into the individual muffin cups. Bake at 350 degrees for about 20 minutes or until they are golden. There's no point in telling you to let them cool just a little bit because if there's anyone near the kitchen, the muffins won't be around that long.

*Crab crossing sign*

# Crab Muffins

These little muffins are perfect finger food for a large or small crowd and can be served warm or at room temperature.

Yields 24 mini muffins

Vegetable spray
¼ cup (½ stick) butter
¼ cup sweet onion, minced
¼ cup celery, minced
2 cloves garlic, pressed
1½ cups strong fish, chicken, or vegetable broth
1 package stuffing mix
1 cup Swiss (or other flavor) cheese, shredded
1 lb. crabmeat
2 tbsp. chives, chopped
3 eggs
¼ cup lemon juice
1 cup milk

Preheat oven to 350 degrees. Coat mini-muffin tins with vegetable spray.

Sauté sweet onion and celery in butter until onion becomes translucent. Add the garlic and sauté an additional 2 minutes. Add broth and bring to a boil.

Remove from heat. Add contents of stuffing mix, gently tossing to combine. Let rest, uncovered, for 15 minutes.

Gently add the cheese, crabmeat, and chives into the stuffing mix.

In a bowl, combine the eggs, lemon juice, and milk until well mixed. Let this mixture rest for 5 to 10 minutes. It will look like it has turned.

Pour egg mixture over crab stuffing mix and toss gently to combine.

Fill the mini-muffin tins short of the rim. Bake for 30 minutes or until golden brown.

# Crab and Cheese Muffins

This recipe uses English muffins, but you could also split crusty French or Italian bread and spread the mixture on halves and broil or bake.

Yields 16 muffins

8 oz. crabmeat
1 stick butter or margarine
1 jar Old English cheese spread
8 oz. cream cheese, softened
1 package English muffins, split

Gently blend together the crabmeat, butter, and cheeses. Spread the mix on the English muffins and place under the broiler for about five minutes or until they've browned. You could also bake in a 350-degree oven for about 10 minutes.

# South City Kitchen Crab Hash

It's difficult to imagine crabmeat being left over, so you may have to break into some crabs or a package of crabmeat to accomplish this recipe from Chip Ulbrich, the award-winning executive chef at South City Kitchen in Atlanta.

The Hollandaise sauce below is easy and fairly foolproof. However, if you're afraid of trying a Hollandaise, buy it prepared from the store.

Serves 4 generously

## Hash

2 large Idaho potatoes
2 large ripe tomatoes
Canola oil
1 lb. jumbo lump crab, picked clean
1 tbsp. butter
¼ cup chives, snipped and divided
8 large eggs
2 cups Hollandaise sauce (recipe follows)

## Hollandaise

3 egg yolks
¼ tsp. Dijon mustard
1 tbsp. lemon juice
½ cup butter, melted

## Hash

Peel potatoes and evenly dice into half-inch cubes. Cover with water and reserve.

Peel and dice tomatoes by first removing the core and cutting a small X on the other end. Blanch quickly in rapidly boiling water, then shock in ice water. Remove the skins and discard. Cut tomatoes across the equator, squeeze out the seeds, then dice into half-inch cubes, reserve.

Prepare Hollandaise and keep in a warm area until dish is ready (recipe follows).

Prepare egg poaching water by bringing some water to a simmer in a wide shallow pan. While the poaching water is heating, prepare hash by heating about ½ inch of canola oil in a large skillet until shimmering but not smoking. Drain and pat dry potatoes and carefully add to oil. Brown potatoes, turning frequently to color on all sides, until they are cooked through. Modulate heat so that they don't scorch. Once cooked, drain well and return to pan.

Add the crab along with 1 tbsp. of butter and brown, tossing frequently. Add diced tomatoes at the last minute to heat through, season with salt and pepper. Toss in half of the chives, remove from heat and reserve.

Crack eggs into simmering water making sure that they are submerged, add more water if needed. Poach to desired doneness.

Divide hash between 4 plates, top each with 2 eggs, then some Hollandaise and the rest of the chives.

## Hollandaise

Blend the egg yolks, mustard, and lemon juice in a blender for about 5 seconds. With the blender on high speed, slowly pour the melted butter into the egg yolk mixture until it is thickened. Do not try to add all the butter at once or at a speed greater than a drizzle.

*Chef Chip Ulbrich, South City Kitchen, Atlanta, Georgia (Courtesy South City Kitchen)*

# Chesapeake Crab Quiche

Quiche and breakfast go together but so does quiche and a late-night snack or card game (real men do eat quiche, particularly if it's made with crabmeat). You could also prepare this dish in individual pie tart shells and serve as a first course at lunch or dinner.

Thanks to the Virginia Marine Products Board in Newport News, Virginia, for providing this delicious recipe.

Serves 6 to 8

½ cup mayonnaise
2 tbsp. all-purpose flour
2 eggs, beaten
½ cup milk
1⅔ cups crabmeat
⅓ cup onion, diced
8 oz. Swiss (or other type) cheese, grated
9 inch pie crust

Blend the mayonnaise, flour, eggs, and milk. Gently stir in the crabmeat, onion, and cheese. Pour mixture into pie crust. Bake at 350 degrees for 30 to 35 minutes or until set.

# Deviled Crab, Breakfast Style

This recipe for individual crab balls calls for mashed potatoes as a filler and slightly crushed corn flakes as breading. I can see switching out with mashed sweet potatoes or garlic potatoes or your favorite mashed potatoes. I can see using almost any type of dry cereal, except, perhaps, for sweetened ones. Panko crumbs would also work.

Other deviled crab recipes are in the entrée section.

Serves 4

2 cups corn flakes, crushed slightly
1½ cups American cheese, grated
1 tbsp. lemon juice
2 tsp. dry mustard
2 tbsp. Worcestershire sauce
2 large eggs, beaten
3 tbsp. green onion, chopped
8 oz. crabmeat
3 cups mashed potatoes

Combine the crushed cereal and grated cheese. Set aside.
Blend the lemon juice, dry mustard, Worcestershire sauce, and eggs.
Add the green onion, crabmeat, and mashed potatoes. Form mixture into small round balls and roll in the corn flake/cheese mixture.
Bake at 375 degrees until golden brown.

# Crab and Apple Beignets

On my first visit to New Orleans, I made the obligatory visit to Café Du Monde for beignets. I even walked around all day with powdered sugar on my blouse as a well-earned trophy. I was so excited when I found a cookbook in the airport with a recipe for beignets that I bought it without thumbing through at all. It wasn't until I was home that I read the recipe that started with "Buy a box of beignet mix . . ."

Chip Ulbrich, an Atlanta chef, explains how to make them, particularly for those of us who don't live in the Big Easy.

Yields 75 pieces

2 cups water
½ lb. butter
¼ tsp. nutmeg
1 tsp. salt
2¼ cups bread flour
8 eggs

In a sauce pan heat water, butter, nutmeg, and salt. Melt butter and bring to a boil, remove from heat.

All at once add flour and stir with a wooden spoon until incorporated.

Return to heat and stir constantly until dough comes away from the sides and is quite smooth, approximately 2 to 3 minutes.

Remove from heat and let rest for 1 to 2 minutes.

Stir in half of the eggs all at once and combine.

Add remaining eggs one at a time, adding the next when the previous is incorporated.

Set aside to cool.

To this mixture add

8 oz. lump crabmeat
1 large Granny Smith apple, very finely diced
¼ cup Parmesan, shredded
1 bunch scallions, sliced
1 tbsp. garlic, chopped
¼ cup parsley, chopped
¼ tsp. cayenne pepper
Salt and pepper to taste

Mix well and drop into 350-degree oil by the small spoonful and cook until golden and light. Fry only a few at a time so they aren't crowded and to prevent the temperature from dropping drastically, which will leave the beignets soggy with oil.

*Crab Pretzels*

# Appetizers and Dips

Individual appetizers made in muffin and mini-muffin tins have an elegant appeal and are neater to eat than the same ingredients on top of a cracker. Using frozen pastry sheets or puffs make them almost as easy to make, too. When using frozen pastry or puff pastry, be sure to thaw before you use them (in the case of phyllo dough, thawing can take up to six hours). Keep them moist, under a damp towel. Return leftover pastry to the freezer or refrigerator or fry little scraps and cover with powdered sugar or a tangy spice for incredibly tasty and easy nibbles.

Because the crab is mixed with other ingredients and this is a matter of taste over appearance, you can use any type of crabmeat or combine several types. Feel free to substitute ingredients, including light mayonnaise for regular, Italian breadcrumbs for panko crumbs, or any similar ingredient that may be more plentiful in the market or preferable to the tastes of your friends and family. I rarely made or make the exact recipe twice so when someone would say he or she loved the tarts or pinwheels or dip I served at a previous gathering and ask for the recipe, I rarely remember what I made. Fortunately, they're all good.

Depending on your will (or won't) power, almost all of these recipes can be made the day before and then heated or brought to room temperature.

Although hot steamed crabs call for cool beer, once you start using crabmeat, you can change your beverage to wine. Red, white, or rosé, it is certainly your choice. Laurie Forster, "The Wine Coach," has two recommendations. The first is Dr. Loosen "Dr. L" Riesling (Mosel, Germany). Laurie says, "This is the perfect white to have on hand for your next gathering because it can pair with a wide range of foods, including crab dip, salads, poultry, and seafood. The secret to its food-friendly abilities is its perfect balance of fruit and acidity with just a touch of sweetness. Retails for $13.00."

Her second choice is Miraval "Pink Floyd" Rosé (Provence, France), saying, "OK, I know this sounds quirky—Pink Floyd and all—but this is a great dry rosé! Back in the day, there was a recording studio on the property where Pink Floyd, Sting, Sade, and others recorded albums. It's made from a blend of Grenaches and Cinsault grapes, and this crisp rosé has flavors of green apple and pear. The Pink Floyd Rosé is perfect for summer favorites like salads, seafood, and yes, you guessed it—crab dip! Retails for $16.00."

*Crab paperweight*

# Crab Tarts

As a frequent hostess in my younger days, I developed a list of easy appetizers and hors d'oeuvres that I could mix and match depending on what I had on hand. With this crab mixture, I could make it a day ahead, adding whatever else sounded good, and then assemble and cook or reheat it before my guests arrived.

Yields 6 tarts

8 oz. crabmeat
4 oz. cream cheese, softened
2 scallions or green onions, thinly sliced
¼ cup mayonnaise
1 package puff pastry shells

Preheat the oven to 400 degrees or follow pastry package instructions.

Gently mix the crabmeat, cheese, onions, and mayonnaise in a medium bowl. Cook over a medium heat until bubbly.

Bake the pastry shells on a baking sheet or as instructed on the package for about 20 minutes or until the pastry shells are light brown.

Remove and reserve the tops from the pastry shells and scoop out the soft interior. Divide the crab mixture among the puff shells and replace the tops.

As an alternative, you can add a tsp. or two of white horseradish or a ¼ cup of slivered almonds to the crab mixture. Another option is to use 2 oz. of cream cheese and 2 oz. of shredded Cheddar or other hard cheese.

# Crab Cheese Spirals

This recipe calls for puff pastry and cream cheese. Another option uses plain white bread that you flatten with a rolling pin and American cheese. Once you've tried them, switch ingredients around, trying different cheeses, until you find the combination you like.

Yields 20 spirals

**Vegetable spray or substitute**
**1 egg**
**4 oz. cream cheese, softened**
**12 oz. crabmeat**
**½ cup water chestnuts, finely chopped**
**¼ cup scallion, thinly sliced**
**¼ tsp. garlic powder**
**½ tsp. crab seasoning**
**1 sheet puff pastry (room temperature)**

Preheat oven to 375 degrees.
Coat a baking sheet with vegetable spray or grease.
Beat 1 egg in a medium-sized bowl and add in the softened cream cheese, water chestnuts, scallion, garlic powder, and crab seasoning, and then gently incorporate the crabmeat.
Unfold pastry sheet onto a lightly floured surface.
Spread the cheese mix onto the pastry. Roll up from the narrow end like a jellyroll, wrap in plastic, and refrigerate for 30 to 60 minutes, until the roll is firm.
Cut the roll into ½-inch slices (about 20) and place on baking sheet, about ½ inch apart.
Bake for 25 minutes or until light brown. Garnish with a sprinkle of crab seasoning.

# Crab Rolls

This is similar to the previous recipe for Crab Cheese Spirals, but uses white bread instead of pastry dough.

Yields 30 rolls

¼ lb. butter, softened
½ lb. American cheese
1 lb. crabmeat
2 loaves plain, white, sliced bread (15-slice loaves), crusts removed
½ cup melted butter

Melt the butter and cheese together in the top of a double boiler, add crab and mix well.

Flatten each piece of bread as thin as possible, spread on the crab mixture and roll.

Cut each small roll in half and chill.

Place flat on a cookie sheet, brush with melted butter, and bake in a 350-degree oven until lightly browned.

*Crab shack sign*

# Stuffed Mushrooms

Button mushrooms are great for this dish and portion size can be easily controlled. I also like shiitake mushrooms for a larger presentation. When I want a more exotic taste, I'll buy a package or two of dehydrated mushrooms in whatever varieties are available at the grocery store. Soak them in warm water for about 15 minutes and then chop and add to the onion and celery mix. These can be made on top of the stove or in the oven if you need the extra burner.

Yields 16 to 20 stuffed mushrooms

1 lb. large button mushrooms (16 to 20)
1 stalk celery
1 onion
6 tbsp. butter, divided
4 oz. crabmeat
½ cup breadcrumbs
½ tsp. crab seasoning blend
Salt and pepper to taste
2 tbsp. Parmesan cheese, grated
Lettuce leaves

Clean mushrooms and remove stems. Chop stems, celery and onion.

Melt 4 tbsp. butter in a skillet and add mushroom stems, celery, and onion, cooking until slightly browned. Gently fold in the crabmeat. Add breadcrumbs and mix until moist. Add seasonings to taste.

Sauté the undersides of the mushroom caps for 1 to 2 minutes. Then turn the mushrooms over and fill with crab/mushroom mix. Sprinkle with a little Parmesan cheese. Cook for about 5 minutes or until the caps are heated through.

Alternatively, you can spray a shallow baking dish (large enough to hold the mushrooms in a single layer) with vegetable spray. Place the mushroom caps in the dish with the bottom side up. Fill each cap with crab/mushroom mix. Sprinkle each mushroom with a little Parmesan cheese.

Or if you have escargot plates, put a stuffed mushroom cap into each snail depression and sprinkle with cheese. This works much better for button mushrooms than for larger mushrooms like shiitakes.

Bake at 350 degrees for 15 to 20 minutes until hot and mushroom caps are tender.

Place lettuce leaves on plate and distribute mushrooms evenly (about 6 per person).

*Chalkboard crab sale sign*

# Mushrooms Chesapeake

Bryan Sullivan, certified executive chef, started with the Admiral Fell Inn, a Harbor Magic Hotel, in 2001 after studying at Johnson and Wales University in Providence, Rhode Island, and working in Ireland and Washington, D.C. (working for Chef Michel Richard at Citronelle). After a two-year stint at Baldwin's Station, Bryan returned to downtown Baltimore as executive chef of Oriole Park at Camden Yards. Trying to satisfy as many as 43,000 people every day is a serious challenge, which he handled very well. He now helps others learn the culinary ways by teaching continuing education classes at Anne Arundel Community College.

Combining local Natty Boh (National Bohemian) beer, Old Bay crab seasoning, and crabs is such a natural it's surprising it hasn't been done before. Bryan has done that and added mushrooms to make this appetizer you'll have plenty of demands for at future gatherings.

Yields 12 mushrooms

1 cup mayonnaise
1 tbsp. lemon juice
2 egg whites
2 tbsp. parsley, chopped
2 tbsp. heavy cream
1½ tsp. Old Bay Seasoning
Salt and pepper to taste
1 lb. jumbo lump crabmeat
12 silver dollar mushrooms
6 oz. smoked Gouda, grated

Mix together the mayonnaise, lemon juice, egg whites, parsley, heavy cream, Old Bay, and salt and pepper.

To make the Crab Imperial, place the crabmeat in a bowl and add enough of the wet mixture to coat the meat (all the wet mixture may not be needed).

Remove the stems from the mushrooms and roast the mushroom caps for 10 minutes at 350 degrees.

Remove the mushrooms from the oven and fill with the Crab Imperial.

Return the mushrooms to the oven and bake until golden brown

Remove the mushrooms from the oven, top them with the Gouda, and return to the oven for 2 minutes to melt the cheese.

# Crab Deviled Eggs

There are three basic options when making and presenting deviled eggs. The first is to combine the crabmeat in the deviled egg mixture. The type of crabmeat used is not important. Use whatever you have or tastes best to you in the mix. The second is to make deviled eggs the way you normally would and top each egg with a sprinkle of crab seasoning and a piece of lump crabmeat. The third is to include crabmeat in the mix and on top.

1 dozen eggs, hard cooked, chilled, and shelled
2 tbsp. mayonnaise
1 tbsp. prepared mustard
½ lb. crabmeat
Salt and pepper to taste
Crab seasoning

Slice eggs lengthwise and remove yolks.
Combine yolks with mayonnaise, mustard, crabmeat, and seasoning.
Spoon or pipe mixture into egg whites.
Top with sprinkle of crab seasoning.

*Steamed-crab shipping boxes*

# Founding Farmers Crab Devil-ish Eggs

Founding Farmers is said to be Washington, D.C.'s greenest restaurant and the first full-service, upscale-casual LEED (Leadership in Energy and Environmental Design) Gold eatery in the United States.

Chef Al Nappo, reared in Tucson, Arizona, started in the restaurant business as so many do, washing dishes for the Tack Room, one of only thirteen 5-star restaurants in the country. Ah, but working for Chef David Benjamin Lolly helped define Nappo's career. After an interesting and fascinating tour of fine eateries, Nappo ended up working for the Cheesecake Factory. Aha, you say, and now you understand why you find such extraordinary takes on ordinary foods there.

This is Founding Farmers take on deviled eggs and when you visit the restaurant to order their devil-ish eggs, you can order a combination (two crab, two Maine lobster, and two salmon) or you can order just one flavor. Note, the yolks of the first set of hard-cooked eggs are not used in this recipe.

½ cup jumbo lump crabmeat
2 tbsp. Founding Farmers Louie Dressing (recipe below)
1 pinch kosher salt, plus more for seasoning
Freshly ground black pepper, plus more for seasoning
8 hard-cooked eggs, halved, yolks discarded, a thin piece sliced off the bottom so the eggs sit flat
Devil-ish Egg Salad (recipe below)
1 tsp. fresh chives, snipped
Old Bay seasoning

In a small bowl, mix together crabmeat and Louie dressing. Season to taste.

Arrange the egg whites on a small platter. Season with salt. Pipe or scoop egg salad mixture into the well of the egg whites. Mound the crab mixture on top, and garnish with chives, Old Bay, and black pepper. Serve immediately

## Founding Farmers Louie Dressing

1 cup mayonnaise
½ cup cocktail sauce
1 oz. sour cream
⅛ cup celery, small diced
½ tbsp. Old Bay seasoning
1 green onion
1 dash Tabasco sauce
¼ tsp. fresh-squeezed lemon juice
1½ tsp. Italian parsley, roughly chopped
1¼ tbsp. sweet relish

In a small bowl, mix together all ingredients until completely incorporated. Any extra can be stored in the refrigerator for up to 3 days.

## Founding Farmers Devil-ish Egg Salad

6 hard-cooked eggs, peeled and diced
2 tbsp. sour cream
½ cup mayonnaise
2 tbsp. yellow onion, diced fine
2 tbsp. celery, diced fine
1½ tsp. fresh chives, snipped
¾ tsp. celery salt
¼ tsp. kosher salt
⅛ tsp. ground white pepper

In a mixing bowl, fold all of the ingredients together and mix well.
Cover and refrigerate until use; unused portion can be refrigerated for up to 2 days.

*Chef Al Nappo (Courtesy Founding Farmers restaurant)*

# Crab and Oysters Bubb-afeller

Travis Todd is a third-generation owner and head chef at Ocean Odyssey on Route 50 in Cambridge, Maryland. This recipe earned him first place in the 2010 Taste of Cambridge competition. He created this dish by adding crabmeat to his normal creamy, garlicky, bacony, semispicy sauce Oysters Bubba-feller. Voilà! Ocean Odyssey restaurant started in 1947 as a crab-processing plant. They moved to their current location in 1986 (yes, they still process crabs in the back) and eventually changed from a strictly carry-out restaurant to a sit-down establishment. Of course, you can substitute whatever oyster is available when you prepare this, but there's no guarantee it will taste as good as the Choptank Sweets.

Incidentally, Travis says it is okay to eat oysters any time of the year. Before adequate refrigeration, oysters were reserved for "months with R in the name," which meant the cooler fall, winter, and spring months. As long as you know where the oysters originated and how they've been transported, you should be fine.

½ lb. bacon
1 tbsp. butter
½ cup onion, diced
1 tbsp. minced garlic
1 tsp. lemon juice
1 pt. heavy cream
2 cups chopped spinach
Old Bay seasoning
Black pepper
Crushed red pepper flakes
⅓ cup dry sherry
12 Choptank Sweets (oysters)
½ cup grated Parmesan cheese
Back fin crabmeat

Chop and cook the bacon; drain and set aside.

Melt the butter in the same pan and add the onion, garlic, and lemon juice. Sauté over medium heat until transparent. Add cream. Stir and allow to reduce until it thickens, about 5 minutes.

Add spinach and stir until it cooks down. Add bacon and the Old

Bay, black pepper, and red pepper flakes to taste. Add sherry and stir all the ingredients together.

Shuck oysters and place a spoon full of sauce over each one. Sprinkle with Parmesan cheese. Grill or broil oysters over high heat until cheese is melted.

Top each oyster with a small clump of back fin crabmeat. Or, you can add the crabmeat to the oysters before you cook them.

*Travis Todd, chef, Ocean Odyssey, Cambridge, Maryland*

# Crab Dip

No matter what else is served, people seem to make a beeline to the crab dip and empty the bowl before they tackle anything else.

This fairly basic crab dip recipe is from Gary Beach, former chef/owner of the Marlin Moon Grille in West Ocean City, Maryland and later the general manager of Micky Fins Dockside Bar & Grill also in West Ocean City. He has received the Restaurant Association's Favorite Restaurant Award and provided me with a handful of tasty tidbits.

1 lb. cream cheese, softened
1 shallot or sweet onion, minced
1 clove garlic, minced
1 heaping tbsp. of sour cream
1 heaping tbsp. of mayonnaise
Juice from ½ fresh lemon
Dash of hot sauce
Dash of Worcestershire sauce
Dash of Old Bay seasoning
1 lb. crabmeat
½ lb. mixed cheeses, shredded (or just American or a cheese you prefer)

Mix the first 9 ingredients and then gently add the crabmeat. Bake at 350 degrees for about 15 minutes or until warm throughout.

Top with cheese and broil until the cheese is bubbly, about 5 minutes.

# Crab Dip, Slightly Spicy

This is a variation for those who want a little bite.

8 oz. cream cheese, softened
8 oz. sour cream
½ cup celery, chopped
3 scallions, chopped or finely sliced
4 tbsp. lemon juice
1 cup Cheddar cheese, shredded
1 lb. crabmeat
Hot pepper sauce to taste (optional)

Blend the cream cheese and sour cream in a medium bowl and mix in the celery, scallions, lemon juice, Cheddar cheese, crab, and hot pepper sauce.

Chill for at least 1 hour before serving. You can heat this to bubbly stage and serve warm.

Use sliced crudités, crackers, cocktail bread slices, or chips with the dip.

*Crab mates*

# Crab Dip with Mushrooms

This dip has spicy and earthy tones so you may want to use a whole wheat or heartier chip or bread, maybe toasted pita triangles to balance the taste.

1 cup sour cream
1 cup creamy salad dressing (such as Ranch)
3 oz. mushrooms, sliced
½ lb. crabmeat
1 cup coconut, flaked (optional)
¼ cup onion, sliced
2 tbsp. curry powder
Salt and pepper to taste

Combine all ingredients and chill. May be heated until bubbly and served warm. Serve with crackers or dipping chips.

# Crab Dip Two

The addition of wine and slivered almonds elevates the sophistication level a little bit.

¼ cup white wine (Chardonnay, Sauvignon Blanc)
½ cup mayonnaise
1 tbsp. Dijon mustard
1½ tsp. powdered sugar
½ tsp. onion juice
2 cloves garlic, pressed
1 lb. crabmeat
½ cup slivered almonds, divided

Combine the first 6 ingredients then gently fold in crabmeat. Add ⅔ of almonds to mixture.

Heat in a 375-degree oven, uncovered, about 35 to 45 minutes or until light brown. Scatter remaining almonds on top of dip and broil for 1 to 3 minutes.

You can substitute a red wine for the white, such as a Shiraz/Syrah, just remember it will change the color of your dip.

*Crab picnic table*

# Founding Farmers Hot Maryland Crab Dip

This is Chef Al Nappo's take on hot crab dip.

½ tbsp. Italian parsley, rough chopped
4 oz. artichoke hearts, drained
4 oz. jumbo lump crabmeat
1 cup mayonnaise
½ tbsp. Parmesan cheese
1 tbsp. yellow onion, diced fine
1 loaf ciabatta bread, sliced and toasted

Preheat the oven to 375 degrees.

In a mixing bowl, thoroughly combine all ingredients. Transfer crab mixture to a shallow casserole dish and place in the oven. Let cook until the sides begin to bubble and the top begins to lightly brown, about 15 to 20 minutes. Serve hot, with toasted ciabatta. The recipe makes about 2 cups of dip.

**Note:** The mix can be held in a covered container and refrigerated for up to 1 day before serving.

# Fab Blue Crab Taboo

Stephanie Holguin, a resident of Northern Virginia, is another online friend who is always fascinating me with her recipes and tales of food. She readily admits that she is "obsessively interested in all things culinary." She is the creator of the popular Web site Foodie-isms.com. Her "lunchbox love" series teaches how to pack "love" and nutrition into a lunchbox (or bag). She says it's for your children, but you're allowed to love yourself, too.

When I asked her for a recipe, she responded in spades. This one doesn't have a detailed story to go with it. She created it for the book.

As she says, "Food is, for many, integral to the experience of a place. That is why it is not surprising that the blue crab is the state symbol for Maryland. The flavors we savor color our journey and play tribute to places once explored. This Fab Blue Crab Taboo appetizer is a sacred tribute to Maryland. Every bite you savor will bring your thoughts right back to the Chesapeake Bay."

**16 oz. of Maryland lump crabmeat**
**⅔ cup mayonnaise**
**2 cloves of garlic, minced**
**2-3 scallions, chopped**
**2 tbsp. prepared horseradish sauce**
**½ tsp. of salt**
**2 lemon wedges**

Add all of the ingredients into a bowl (except the lemon wedges) and stir gently to incorporate without breaking up the crabmeat lumps. Squeeze in the juice of 2 lemon wedges and stir. Keep refrigerated until ready to serve. Then, put in a bowl surrounded by crackers or in the center of a lettuce bed on appetizer plates.

*Fab Blue Crab Taboo (Photograph by Alex Vasilescu)*

# Crab Dip for a Group

This is an excellent recipe if you're entertaining a large crowd.

24 oz. cream cheese, softened
½ cup mayonnaise
¼ cup dry white wine
2 tbsp. Dijon mustard
1½ tsp. sugar, powdered or granulated
½ tsp. onion juice
2 cloves garlic, pressed
1 lb. crabmeat
½ cup toasted slivered almonds
¼ cup minced fresh parsley

Combine first 7 ingredients.

Fold in crabmeat and cook in a 350-degree oven for about 20 to 30 minutes or until bubbly. You can mix everything and chill overnight before heating prior to serving. Add a little heating time if you're cooking from a refrigerator temperature.

Transfer to warm chafing dish, sprinkle with almonds and parsley.

# Crab Dip, Creamier Style

3 hard-cooked eggs, peeled and finely chopped
1 cup evaporated milk
1 cup mayonnaise
1 tsp. onion, grated or finely chopped
½ tsp. salt, or to taste
½ tsp. pepper, or to taste
8 oz. water chestnuts, drained and finely chopped
1½ cups breadcrumbs, divided
8 oz. crabmeat
1 tbsp. butter

Combine the first 7 ingredients, including 1 cup of the breadcrumbs. Then add the crabmeat.

Place in a baking dish, top with remaining breadcrumbs and dabs of butter. Bake at 350 degrees for 30 minutes.

Mixture may be made the day before to allow flavors to meld. Bake just before serving.

Serve in a bowl surrounded by crackers or chips.

*Fresh Maryland crabs sign*

# Crab Pretzel

I'll admit to being late to this taste craze, but it's so logical a combination that I can't imagine that it hasn't been around for decades. I've heard restaurants in Annapolis, Baltimore, Boston, and Philadelphia claim rights to the best ever. Chops for the original or first salty offering seem to go to the Silver Spring Mining Company in Silver Spring, Maryland. They have other locations in Bel Air, Hunt Valley, and Perry Hall.

You can experiment with the crab mixture and then play around with the best method to pipe the mixture on top of the pretzel. You could just try to spoon it on. We should all have so much fun.

Yields 8 pretzels

3 oz. cream cheese, softened
1 tbsp. crab seasoning
2 tsp. onion, minced
1 tbsp. hot sauce (optional)
1 lb. crabmeat
Large soft pretzels (preferably unsalted)
½ lb. Cheddar cheese, shredded
½ lb. Jack cheese, shredded

Combine the cream cheese, seasoning, onion, and hot sauce in a large bowl, and then gently fold in the crabmeat. Either spoon the mixture on top of the pretzel or put the mixture into a piping bag, a large plastic bag with the tip cut off, or a cookie press and pipe the mixture over the pretzel in about a ½-inch bead. Mix the cheeses and sprinkle on top of the crab mixture. Place pretzels on baking dish lined with foil or a Silpat to catch any drips and make cleaning up easier. Broil or bake in a 350-degree oven until cheese has started melting.

# Crab Pretzel Two

You can use regular cream cheese or other flavor (sun-dried tomato and basil would be a good place to start) instead of the chive and onion. You can add a couple of tablespoons of crumbled bacon or bacon bits. Most recommendations seem to favor a sweet/sour or honey mustard sauce for dipping, just in case the pretzel isn't enough.

Yields 6 pretzels

3 oz. chive and onion cream cheese, softened
½ cup mayonnaise
1 tbsp. white horseradish
½ tsp. garlic, minced
2 tbsp. onion, minced
1 tbsp. lemon juice
½ tsp. hot pepper sauce (optional)
½ lb. crabmeat
White pepper to taste
Soft pretzels
2 cups shredded cheese (mixture of your favorites, from Cheddar to Colby)

Preheat oven to 350 degrees.

Blend cream cheese, mayonnaise, and horseradish in a large bowl until smooth.

Stir in the garlic, onion, lemon juice, and hot pepper sauce. Gently fold in the crabmeat, and season with white pepper to taste. Spoon or pipe mixture evenly over the pretzels and sprinkle each with ½ cup shredded cheese.

Place pretzels in the preheated oven until the cheese is melted and the topping is bubbly, about 20 minutes.

# Crab Cocktail

I think one of the first serving dishes I bought was a set of nested shrimp cocktail dishes where you put crushed or shaved ice in the bottom and torn lettuce, shrimp, and cocktail sauce in the top bowl. Oh, what the dishwasher did to those top bowls. Are there any left anywhere or did the Tavern on the Green auction sell the last of them?

I like a creative presentation for a crabmeat cocktail so I suggest Cosmo glasses. No, you don't have to go out and buy a set. Martini or margarita glasses will do. In fact, regular appetizer plates will be just fine. I suggest the glasses because you can dip the rims in lemon juice and then crab seasoning (like a margarita or any other drink with a frosted rim). Short of the special glasses just dust the top of the dish with a sprinkling of crab seasoning.

**Note:** Zest the lemon before you juice it.

Serves 4

1 tbsp. mayonnaise
2 lemons, zest grated and juiced, plus juice for coating glass rims
1 tbsp. Dijon mustard
1 lb. crabmeat, back fin or regular lump
Salt and pepper to taste
Crab seasoning
4 lettuce beds, endive, Romaine, or iceberg
Lemon slices, for garnish

Mix the mayonnaise, lemon zest and juice, and mustard and then gently incorporate the crabmeat. Salt and pepper to taste.
Dip the glass rims in lemon juice and then into crab seasoning.
Line each glass with lettuce and fill with the crab mixture.
Garnish with lemon slices. Serve with or without crackers.

# Crab Meltaways

Earlier in this book, I mentioned David DeBoy, the actor, writer, performer, director, producer, and creator of the Baltimore holiday hit song *Crabs for Christmas*. Other than his crab-shaped cookies (see the dessert section), he proudly acknowledges that it's his wife, Joellen DeBoy, who is the chef, and she was nice enough to contribute two recipes to this book. This recipe is a favorite of Helen Manik, Joellen's mom, who has been preparing meltaways for more than twenty years.

Yields 96 portions

1 lb. crabmeat
2 jars Kraft Old English Cheddar Cheese
2 sticks margarine
4 tbsp. mayonnaise
1 tsp. seasoned salt
1 tsp. garlic salt
1 dash of Tabasco sauce
12 English muffins, sliced in half and quartered

Mix first 7 ingredients in mixing bowl. Place portion of mixture to cover top of each muffin quarter. Bake on baking sheet at 350 degrees for 12 to 15 minutes or until golden brown. Halve recipe for smaller yield. The prepared unbaked meltaways can be frozen on sheet in freezer then placed in storage bags to be baked later.

# Joe's Crabbies

Thanks to the Internet, I have friends all over the world. Dick Schock is an online friend I've never met in person (or IRL—in real life—as "they" say). I know he lives up north someplace and he'd love to open a restaurant or tavern somewhere on Virginia's Eastern Shore so I'm guessing we'll meet up one day. In the meantime, I know he's collected about two hundred soup recipes for that tavern so I asked him for something featuring crab.

Instead, he sent this recipe for Joe's Crabbies. It's not soup, but it is a variation on the crab meltaways that Joellen DeBoy prepares.

Yields 30 pieces

1 stick sweet, unsalted butter
1 jar Kraft Old English cheese spread, at room temperature
1½ tsp. mayonnaise
½ tsp. garlic salt
½ tsp. salt
7 oz. can crabmeat
1 package Thomas's English muffins or bagel chips
Old Bay seasoning

Melt the butter in a bowl and then stir in the cheese spread. Add the mayonnaise and salts. Gently fold in the crabmeat.
Split the muffins in half and then quarter each half. Spread the mix onto the muffin pieces and freeze for at least 10 minutes.
Broil until bubbly and crisp. Sprinkle with a little Old Bay seasoning and serve warm.

# Jumbo Lump Crab Bruschetta

Gary Beach says this recipe is an example of "necessity being the mother of invention. At the Marlin Moon Grille we found an awesome product, fire-roasted red and yellow tomatoes; however, we were having a hard time figuring out what to do with them. We also had Brie and fresh jumbo lump crabmeat we had to move. We put them all together and this is what happened. It's nice, simple, and easy to prepare. You can also use par-grilled, halved jumbo shrimp; grilled chicken strips; Italian sausage; or it's great without seafood," he says.

8 slices crusty bread, cut in ⅓ inch slices
Extra-virgin olive oil
Salt and pepper, to taste
8 slices Brie or triple cream cheese
8 pieces each roasted red and yellow tomatoes (check your market's olive bar)
4 oz. fresh Maryland J.M. Clayton Company "Epicure" Jumbo Lump Crabmeat
Chiffonade of fresh basil

Brush sliced bread with extra-virgin olive oil and apply salt and pepper. Toast for 3 to 5 minutes in a 350-degree oven or grill lightly. Top each piece with 1 slice of cheese, then roasted tomatoes, and finally crabmeat. Bake in 350-degree oven until toasty. Top with basil and a light squirt of extra-virgin olive oil.

**Note:** To chiffonade basil, stack fresh basil leaves on top of each other, roll, and slice end to end.

*Boat-shaped crab condiment trays*

# Crab Wontons in Creamy Shallot Sauce

Makes 24 wontons

### Stuffing

12 oz. lump crabmeat, or 2 6-oz. cans, reserve liquid
1 cup ricotta cheese
1 tsp. oregano
2 packages wonton wrappers (not shrimp roll papers)

### Sauce

1 tbsp. olive oil
1 shallot, chopped fine
½ cup dry white wine
1 cup chicken or fish broth or water, or add liquid from the crabmeat cans
½ pint heavy cream
1 tbsp. parsley, chopped fine
Salt and pepper to taste
Parmigiano-Reggiano, shredded

### Stuffing

Mix the crabmeat with the ricotta and oregano.

Put 1 tsp. of the mixture in the middle of the wonton, dampen the edges, and fold over to seal, pressing firmly.

Just before serving, place wontons in boiling water. Do not overcrowd pot. Cook for 5 minutes (wontons will rise when cooked), then remove and place on a cookie rack to cool and drain. Repeat until all the wontons are cooked.

### Sauce

Heat olive oil in a skillet and sauté shallot until tender. Add wine, broth, and cream and simmer until thickened. Add parsley and salt and pepper to taste. The sauce will need less salt if broth contains salt.

Ladle sauce over wontons, then sprinkle with Parmigiano-Reggiano desired.

# Crab Spread

This is another version of a crab dip, although it's a little thicker than the dip and can be served in a chafing dish or spread on toast points, Melba toast, or similar bread or cracker.

½ lb. crabmeat
8 oz. cream cheese, softened
1 tbsp. milk
2 tsp. Worcestershire sauce
2 tbsp. green onion, chopped
Dash red pepper
Paprika
2 tbsp. slivered almonds (optional)
Parsley

Combine crabmeat, cream cheese, milk, Worcestershire, and green onion. Season to taste with red pepper and paprika. Pour into greased shallow baking dish.

Bake at 350 degrees for 15 minutes or until bubbly. Sprinkle with paprika and parsley or top with almonds. Serve in chafing dish.

*Floor tiles*

# Crab Spring Rolls

This appetizer is a great way to use fresh vegetables from your garden or the farmers' market. It serves double duty because you can use spring rolls for a lighter dish and egg roll wrappers for a heartier one.

Yields 12 spring rolls

½ lb. crabmeat, claw or special
2 tsp. soy sauce
Salt and freshly ground white pepper
¾ tsp. sesame oil
3 dried shiitake mushrooms, reconstituted with warm or boiling water
2 tbsp. vegetable oil, plus oil for deep-frying
1 clove garlic, minced
1 small carrot, julienned
½ cup jicama, julienned
½ cup fresh green beans, julienned
3 large shallots, thinly sliced (not minced)
1 cucumber, seeded and julienned, or English or seedless cucumber
12 spring roll wrappers, thawed if frozen (see Note)
1 beaten egg
Sweet and sour sauce or duck sauce

Mix the crabmeat with the soy sauce, salt and pepper, and sesame oil in a container with a lid. Marinate and refrigerate for at least 30 minutes.

Remove the stems of the shiitake mushrooms and thinly slice the caps.

Heat the vegetable oil in a pan (a wok works well) and sauté the garlic for about 2 minutes, then add the carrot, jicama, green beans, shallots, cucumber, and mushrooms and stir fry until the vegetables are slightly wilted. Remove vegetables from the pan and drain on paper towel.

Take a spring roll wrapper and wipe the edges with some of the beaten egg. Place about ¼ cup of the filling diagonally on the wrapper and roll from one corner, tucking in the edges (deli wrap). Fill the other wrappers.

Heat the vegetable oil to 375 degrees and fry the rolls, turning once, for about 2 minutes or until lightly brown and crispy. Cook only a couple of rolls at a time to avoid crowding and a sudden drop in temperature. Drain on a paper towel and then put rolls in a warm (250 degrees) oven until they are all cooked.

For a pretty presentation, slice each roll in half diagonally and place on a bed of lettuce on a serving tray. Put sweet and sour sauce or duck sauce in one or more dipping bowls. Or serve on individual appetizer plates.

**Note:** If you're using egg roll wrappers, cook for another minute or two until the rolls are golden brown.

*Wall crabs*

# Crab Dumplings

I traveled to Norfolk, Virginia, in 2010 to research a couple of articles and, of course, I had to find someplace to eat. Erin Filarecki, media relations manager for Visit Norfolk, showed me around town. My mother's family is from the Tidewater area, so I love returning, seeing what's new and what's changed, and so much of Norfolk has changed for the better.

While we were lunching at 219 Bistro, we talked to Joe Haggard and I asked him for a crab recipe. He went upstairs, and I guess talked to executive chef Amarin Reelachart, and returned with this recipe, handwritten. For a restaurant that had just opened in May of 2010, they were doing great business and offering a solid and interesting menu. I just wish the crab dumplings had been on the menu that day. I guess they would have been a little heavy for a hot summer afternoon.

Yields 24 dumplings

## Filling

1 lb. lump crabmeat
2 tbsp. scallion, chopped
½ tsp. ginger, minced
¼ cup water chestnuts, chopped
1 tbsp. soy sauce
½ tsp. sugar
1 egg yolk, beaten
2 tbsp. cornstarch
24 pieces wonton or dumpling skins

## Dipping Sauce

1 cup soy sauce
2 tbsp. sugar
1 tsp. garlic, minced
1 tbsp. sesame oil
1 tbsp. scallion, chopped
2 tbsp. white vinegar

### Filling

Mix all ingredients except wonton skins in a medium bowl and mix gently but well. Chill for 45 minutes.

Lay out wonton or dumpling skin on flat surface. Place ½ to 1 oz. of crab stuffing in center of skin. Lightly pat water around the edge of the skin. Bring all the edges together to meet in center of filling and pinch firmly to hold into place. Repeat until filling is used. May be made 1 day ahead and frozen until use.

Use a small bamboo steamer lined with parchment paper. Place one level of dumplings in steamer spread apart for hot steam to cook. Cover and place over hard boiling water. Cook 10 minutes. Remove, plate, and serve right away with dipping sauce.

### Dipping Sauce

Combine all ingredients in a medium-sized mixing bowl and mix well. Sauce can be kept in an airtight container in the refrigerator for 1 week.

*Crab crackers and mallets*

# Crab and Shrimp Ceviche with Watermelon Relish

Chip Ulbrich is the executive chef at South City Kitchen in Atlanta where he combines a Southern approach to food with his European-style culinary apprenticeship. This award-winning chef has been generous enough to send along this recipe with a Charleston, South Carolina, touch.

Relish

4 cups seedless watermelon, finely diced
½ cup red onion, finely diced
1 cup red bell pepper, finely diced
1 jalapeño or poblano chile (depending on the spice level desired), finely diced
2 thumb-sized pieces of ginger, peeled and grated
2 limes, juiced and zested
¼ cup champagne vinegar
¼ cup garlic, chopped
1 cup extra-virgin olive oil
2 tsp. ground cumin
2 tsp. ground coriander
¼ cup honey
Salt and pepper, to taste

2 lb. fresh shrimp, peeled and deveined
1 lb. lump crabmeat, picked clean

Mix all ingredients for relish and allow to rest in the refrigerator for a few hours, or overnight.

To blanch shrimp, bring a large pot of water to a rolling boil and season with a healthy dose of salt. The water should taste like seawater. Add shrimp and cook for no more than 10 seconds. Remove the shrimp and shock in ice water to cool rapidly; shrimp should still be a little raw on the inside. Once cool, drain well.

To assemble toss shrimp, crab, and relish in a bowl and allow to marinate for at least 10 minutes but no more than 30 minutes. Serve with fresh watermelon wedges and lime wedges.

# Spider Sushi Maki Roll

Although it's not difficult, this recipe is easily the most complex of the appetizers, so it's the last one in this section. It is time-consuming, but oh, so delicious. With the tangy vinegar sticky rice, the tempura-fried soft-shell crab, the soft little pop of fish roe or caviar, so many things are happening in your mouth at the same time that it is worth the time and effort.

The first time I had a spider roll was at a restaurant at the Gaylord Opryland resort in Nashville. It was damaged in the 2010 flood and has returned as Wasabi's in the Cascade's lobby.

Unless you make sushi regularly, you're probably going to need some items that aren't in your pantry. You will need nori (seaweed sheets), sushi rice (different from regular rice), and some practice in rolling the ingredients. As you become more proficient in making these rolls, you can make them with the rice side out with sesame seeds, with sliced crab legs or sticks (surimi), or other ingredients.

### Rice

1 cup Japanese sushi rice
½ cup water
¼ cup seasoned rice vinegar (see Note)

### Tempura Batter

1 egg, well beaten
1 cup ice-cold water
1 cup all-purpose flour
1 tsp. salt
Canola oil
4-6 soft-shelled crabs

### Spider Roll Assembly

Rolling mat
Nori sheets, cut in half
1 fried soft-shell crab
¼ cup vinegar water
2 cups sushi rice
¼ cucumber, sliced lengthwise
Sesame seeds
1 oz. red caviar (optional)

### Rice

Rinse the rice with cold water until the water is clear. Let the rice drain in a colander and sit for 30 minutes.

Place the rice in a heavy-bottomed pot and add 2¼ cups water and simmer, covered, for 10 minutes. Remove from the heat and let sit, covered, for another 10 minutes.

Spread the rice over a flat surface (a Silpat in an edge cookie sheet is good) and drizzle the seasoned rice vinegar over the rice, 1 to 2 tbsp. at a time, fluffing the rice and fanning it (newspaper, file folder, electric fan) to help it cool as you do. The rice will be sticky enough to hold its shape but have a light, airy texture that falls apart when you bite into it.

**Note:** You can make your own vinegar with ¼ cup regular rice vinegar into which you've dissolved 3 tbsp. sugar and 1 tsp. salt

### Tempura Batter

Put the well-beaten egg and ice water into a bowl. Slowly add the flour until lightly mixed. Add salt to taste.

Heat about three inches of canola oil in a wok. Drop a little tempura batter into the oil and if it rises (floats), the oil is hot enough. Batter the crabs. Cooking them one at a time, hold each one with chop sticks or tongs and dip into the hot oil for a few seconds before releasing. Fry for 2 to 3 minutes or until golden brown. Remove the crabs and drain on a paper towel.

### Spider Roll Assembly

Cover the rolling mat with plastic wrap to prevent the rice from sticking to it. Put the shiny side of the Nori on the rolling mat, long side facing you.

Cut the crab into six pieces, so they look sort of like julienned slices.

Dip your hands in the vinegar water and shake off excess, so the rice won't stick to you.

Take about ½ cup of the sushi rice and pat gently into an even layer about ¼ inch thick on the Nori, just short of the top (the part farthest away from you) by about 1 inch. Flip the rice and Nori over so the Nori sheet is on top.

Place 2 crab pieces, side by side with legs toward the edge, and 2 or 3 cucumber strips, side by side, across the mat, at the edge closest to you. Let the legs hang out of the roll a little bit.

Pick up the rolling mat at the edge closest to you and slowly start rolling very tightly away from you. Gently prod or poke the ingredients into the right place, if necessary. When the end of the mat touches the rest of the rice, squeeze the roll tightly and then release the edge of the mat so it won't be in the roll. Continue rolling the rest of the rice. Squeeze the roll and then make sure the edge of the Nori without the rice is sealed against the roll. If it doesn't seal, dampen your fingers a little and run down the edge of the Nori to help it seal.

Remove the roll from the mat, roll in sesame seeds, and place on a cutting board. Dampen a sharp knife (sushi knife if you have one) and cut the roll in half. Place the 2 pieces next to each other and cut in half again and then repeat until you have 8 pieces of even sizes. Top with a dab of red caviar.

*Spider roll*

*Legal Sea Foods Crab Cakes (Courtesy Legal Sea Foods)*

# Crab Cakes

As far as I'm concerned, there is only one recipe and only one way to cook crab cakes. My recipe is listed under the *Maryland Lady* Crab Cakes recipe. However, if I only acknowledge one recipe, this would be a very thin cookbook. So find a recipe you like or start with one and modify to your heart's content.

Try using all lump crabmeat or a mixture of crabmeat. You can also experiment with the breadcrumbs; try using Japanese panko, chive panko, Italian, plain, or whole wheat. Some people swear by crumbled stale (or fresh) bread and some by Saltines. If you are gluten intolerant, look for chickpea flour, rice breadcrumbs, or brown rice breadcrumbs.

Once the mixture is complete, put it in the refrigerator for an hour or more (if you can possibly wait that long). Form the mixture into cakes (about four ounces or whatever will fit in your hand) or tiny bites for appetizers and broil (or grill) for about five minutes on each side until golden and crispy on the outside. Remember, the crabmeat has been cooked already, so you're only heating it through.

Beyond the various recipes, crab cakes can be separated into three categories. The first, and the best, are crab cakes that you make yourself. You control what kind of crabmeat, how much breading and other ingredients, and whether they're broiled (as they should be), sautéed, or fried (oh, well, everything fried tastes good).

The second are restaurant crab cakes. Chef Michael Symon compared the Faidley's crab cake to Gertrude's at the Baltimore Museum of Art on his Food Network show *Food Feuds*. Finding the "best" crab cake in Baltimore or Maryland is a regular battle. Duff Goldman, of Charm City Cakes and a regular on Food Network shows, says his favorite is at Pierpoint Restaurant on Aliceanna Street in Baltimore, where Nancy Longo smokes the crabmeat first so it has a caramelized smokiness to it. She forms the crab cakes and then sautés them. My favorite restaurant crab cake is served at G & M Restaurant in Linthicum. It is just so much fun trying to decide which you like best.

The third are mail-order crab cakes. The Crab Cake Guy conducted an unscientific study to determine the best place to order delivery crab cakes in the summer of 2010. Judging on a number of criteria, he compared Crab Cake Express, Faidley's Crab Cakes (in Lexington Market, Baltimore), Crab Place, Phillips, Authentic Maryland Crab Cakes, Timbuktu Crab Cakes, the Narrows Restaurant Crab Cakes, the Crab Cake Lady, Koco's Pub, and Cape May Crab Cakes. Some offered online ordering while others, still in the dark ages, only offered ordering by telephone. As the anonymous Crab Cake Guy says, "The delivery times on the crab cakes ranged from overnight to second-day mail. For the most part each delivery crab cake was well prepared, packaged, and the reheating process didn't take away from the taste."

The best delivery establishments are Narrows Restaurant on Kent Narrows, Timbuktu Crab Cakes in Hanover, and Phillips in Baltimore, Maryland. The runner-up was Koco's Pub in Baltimore. These are all Maryland businesses.

The Crab Cake Guy agrees, though, that the best crab cakes are home made. I think there are as many variations as there are crabs in the Bay.

Now, if you're from the Ybor City area of Tampa, Florida, or if you've visited there, you may have come across something that looks a little bit like our crab cakes. They're called "devil" crabs or croqueta de jaiba, but they are more like a croquette in a slightly ovoid shape than what we consider deviled crab. The main difference is the Ybor City version uses a tomato-based mixture rather than a mayonnaise base. Additionally, you can pick up a croqueta de jaiba and eat it without a bun or fork. That's not the case with the flaky, delicate lack of cohesion of a Bay crab cake.

# *Maryland Lady* Crab Cakes

This is the first crab cake recipe I ever had and it came from the chef of the *Maryland Lady*, the state yacht back in the 1980s or thereabout. I think he was also the chef at the Governor's Mansion, but I've forgotten his name and I haven't found anyone who remembers. I have since replaced the Italian breadcrumbs with Japanese panko. Harry Mudrick, my late father, never said much about my cooking (good or bad), but he would devour whatever crab cakes I made and he never complained.

Yields 4 crab cakes or 16 crab balls

½ cup Italian breadcrumbs
1 large egg
1 tbsp. mayonnaise
1 tsp. seafood seasoning
¼ tsp. white pepper
1 tbsp. dry mustard
1 lb. crabmeat, lump or back fin

Mix all ingredients together except the crabmeat and then add it gently. Shape into 6 crab cakes. Refrigerate at least 1 hour. Place on aluminum-foil-lined baking sheet. Broil, about 4 inches from flame or element, on each side, about 5 minutes each until golden brown.

If you're planning to serve crab balls (about 1 inch in diameter), you probably will do better if you sauté them so you can watch how they're progressing and browning.

Drain on paper towels and serve alone or with a sauce.

# Stinnett Family Crab Cakes

Joyce Stinnett Baki, tourism specialist for Calvert County Department of Economic Development, found this recipe in a very old cookbook done by a local church and credited to Elizabeth Stinnett, so Joyce contacted Wesley Donovan (Elizabeth's great-grandson) for permission to use the recipe, and he replied "CRABsolutely!" Then she checked with Ethel Lou (Aunt Lou), Elizabeth's only surviving daughter, who also gave her permission. Elizabeth was born in 1901 and loved fishing the Bay. She was a frequent hostess on fishing trips for her family and friends. On her eightieth birthday, the late Honorable Louis L. Goldstein (known for saying "God Bless You ALL") proclaimed her "Admiral of the Chesapeake Bay."

Yields 4 crab cakes

1 egg
1 tbsp. mayonnaise
Flip of Worcestershire
1 tbsp. baking powder
Squirt mustard
2 slices bread, crust removed
1 lb. crabmeat, back fin
1 tsp. Old Bay seasoning
Pinch parsley

Mix the egg, mayonnaise, Worcestershire sauce, baking powder, and mustard in a bowl.

Tear bread into small pieces and add to mixture.

Mix back fin crab, Old Bay, and parsley, mixing lightly. You do not want to break up the lumps of crab. Add to mixture and toss lightly until it holds together.

Deep-fat fry or broil until golden brown.

## University of Delaware, Best Crab Cake in Delaware

One of the most enduring crab cake competitions takes place during the annual Coast Day at the University of Delaware's College of Earth, Ocean, and Environment, in Lewes, Delaware. The contest for Delaware's best crab cakes is one of the most popular of the day's festivities. The contestants and their temporary kitchens ring a large tent, and bystanders, who have a chance to taste the goodies, watch them prepare their dishes.

Even more important, the competitors for the 2010 contest, the thirty-fourth annual Coast Day, had to please three very discerning judges, Peter Mandelstam (president of Bluewater Wind, LLC), Raymond Williams of Bear, Delaware (2009 winner and Celebrity Kitchens chef), and Karen Falk (Coastal Cakes in Rehoboth Beach).

Some of the previous winning recipes have included Caribbean jerk spice, heavy cream, Frangelico liqueur, crushed Rice Krispies, cornmeal, peach preserves, seafood mousse, and salsa.

*Evan Williams, left, assists his dad Raymond Williams (2009 Coast Day winner) at the judging table at the 2010 Coast Day crab cake competition, University of Delaware. Karen Falk and Peter Mandelstam were the other two judges.*

# Calypso Crab Cakes

George "Geo" Johnson, of Frankford, Delaware, took second place at the 2003 Coast Day crab cake competition. As he was the executive chef at the Cottage Café in Bethany Beach, he could have had a slight advantage. Oh, and he was a judge the year before because he was victorious the year before that. This recipe has lots of crabmeat infused with roasted red and green peppers and then topped with chipotle sauce. Then, Johnson serves the crab cake with a roasted soft-shell crab on a bed of crisp greens.

Serves 8

## Presentation
8 large soft-shell blue crabs (live, if possible)

## Crab Cakes
¼ cup roasted red bell pepper, diced
¼ cup roasted green bell pepper, diced
2 eggs, beaten
4 tbsp. mayonnaise
4 tbsp. Worcestershire sauce
2 tbsp. fresh parsley, chopped
1 tsp. dry mustard
2 tsp. Old Bay seasoning
2 lb. crabmeat, jumbo lump
½ cup cracker crumbs

## Chipotle Sauce
1 egg, beaten
1 tbsp. chipotle base
2 tbsp. lemon juice
1 tbsp. Worcestershire sauce
1½ cups mayonnaise
1½ tbsp. Old Bay seasoning
1 tbsp. dry mustard
Lemon wedges

## Crab Cakes

Before making crab cakes, wash and clean soft-shell crabs, using scissors to snip around the sides of the dark top of the shell. Gently remove and discard. Set aside.

Roast 1 red and 1green pepper; dice. In a mixing bowl, whisk together eggs, mayonnaise, Worcestershire sauce, peppers, parsley, mustard, and Old Bay seasoning. Add crabmeat and cracker crumbs. Toss, taking care to keep meat in chunks, until evenly combined. Form into patties (approximately 2 oz. each).

## Chipotle Sauce

In a separate mixing bowl, whisk together all ingredients until smooth.

## Preparation

Preheat oven or broiler to 450 degrees. On greased sheet pan, place soft-shell blue crabs (backside up) and top each with 2 tbsp. of chipotle sauce. Place 2 crab cake patties beside each crab. Roast for 6 minutes until crab cakes are golden brown.

Place 1 crab cake patty on a bed of crisp greens. Top with 1 cooked soft-shell crab, sauce side up, and top with another crab cake patty. Press gently together to keep stacked. Serve with lemon wedge.

*Crabs by the bushel*

# Pan-Seared Blue Crab Cakes with Pepper and Onion Jam

Steven M. Ruiz, of Wilmington, Delaware, made the best crab cake in Delaware in 2010. Ruiz is the chef de cuisine at Maris Grove, a retirement community in Glen Mills, Pennsylvania, and this was his third attempt at the competition, finishing second last year. His first place prize, besides bragging rights, was $200 and a sterling silver serving plate.

Serves 8

1½ cups mayonnaise
2 eggs
2 tbsp. Old Bay seasoning
1 tbsp. dry mustard
¼ cup lemon juice
1 tsp. Worcestershire sauce
1 tsp. Sriracha hot sauce
2 lb. crabmeat, lump
1 cup Japanese panko
1 tsp. each salt and pepper
1 cup olive oil
1 cup flour

## Pepper and Onion Jam

2 red peppers, diced
1 jalapeño, diced
1 small white onion, diced
4 tbsp. red-wine vinegar
¼ cup red currant jam

In a medium bowl, place mayonnaise, eggs, Old Bay seasoning, dry mustard, lemon juice, Worcestershire sauce, and Sriracha hot sauce. Mix well.

In a separate bowl, take crabmeat and slowly add the wet mix and ½ cup of the panko. Fold together lightly then pat out 4 oz. cakes with that mix. Then, with the remainder of panko, lightly bread the outside of each cake. Sauté in olive oil until crispy and golden brown.

## Pepper and Onion Jam

Sauté the peppers and onion until caramelized, deglaze with red-wine vinegar, and stir in a ¼ cup of the red current jam. Serve with crab cakes.

*Crab spice bags*

# Sussex County Low Country Crab Cakes with Crabanero Remoulade Sauce

Charles Parkhill, of Millsboro, Delaware, placed second in the 2010 Coast Day best crab cake competition, winning $150 in cash. The tasty addition that played such an important part of this recipe is the remoulade. This is also the first recipe I've come across that calls for matzo meal.

Yields 8 crab cakes

1 tbsp. butter, divided
¼ cup celery, finely diced
¼ cup red and yellow bell peppers, finely diced
1 tbsp. Worcestershire sauce
2 tbsp. coarse mustard
¼ cup mayonnaise
1 egg plus 1 yolk
1 tbsp. Crabanero brand hot sauce
¼ cup scallion, shredded and divided
1½ tbsp. Crabanero Bay* Seasoning
2 lb. crabmeat, lump
2 tbsp. parsley, minced
½ cup matzo meal
1 cup Japanese panko
Lemon slices

## Crabanero Remoulade

1 cup mayonnaise
½ cup ketchup
¼ cup celery, minced
¼ cup red onion, minced
¼ cup dill pickle, minced
1 tbsp. smoked paprika
2 tbsp. course ground mustard
1 tbsp. Crabanero brand hot sauce
1 tbsp. minced capers

Sauté peppers and celery in half the butter until wilted. Turn off the heat and add Worcestershire sauce and coarse mustard. Toss, set aside, and allow to cool before proceeding with remainder of recipe.

Blend pepper mixture with mayonnaise, 1 egg plus 1 yolk, Crabanero, and half the scallion and bay seasoning. Pick through crabmeat for shell or cartilage, gently toss crabmeat with parsley, bay seasoning, and enough matzo meal to absorb some of the crab liquid (you may not use all matzo meal, you don't want to be accused of a crab cake with too much filler). Next add the pepper mayonnaise mixture gently.

Allow to stand in cooler before you pat out the crab cakes. Portion crab cakes out, forming into 3 to 4 oz. cylinders.

Toss panko crumbs with remaining scallions, scatter on plate or sheet pan, and place cakes on plate to coat with crumbs. GENTLY press remainder of scallion crumbs into flat top of crab cakes. Add remaining butter to a medium-high heat skillet (do not crowd). Brown one side then the next. Put into oven at 350 degrees for 10 to 15 minutes.

## Crabanero Remoulade

Combine all ingredients well, set aside (best if chilled for 15 to 20 minutes). Garnish plate with lemon, 1 crab cake, and remoulade sauce.

*Crabanero Bay is the original bay seasoned habanero sauce located in Millsboro, Delaware.

*Crab welcome sign*

# Chesapeake Crab Cake

Keith Starkey and Joe Joyce, of Wilmington, Delaware, came in third in the best crab cake competition at the Coast Day festivities. However, they come in first for the fewest ingredients and shortest instructions, and the first one, I think, to incorporate allspice. They took home $100 for their third-place win.

2 tbsp. yellow mustard
⅛ tsp. allspice
⅛ tsp. Worcestershire sauce
⅛ tsp. hot sauce
⅛ tsp. Old Bay
1 cup mayonnaise
⅔ cup Japenese panko, firm to touch
2 lb. crabmeat

Combine mustard, allspice, Worcestershire sauce, hot sauce, Old Bay, and mayonnaise. Toss lightly. Add panko and crabmeat and form crab cake. Broil in the oven for 5 to 10 minutes or until golden brown, then turn to broil the other side for about 5 minutes.

*Crab suncatcher*

# Harbor Magic Maryland Crab Cake

This is Bryan Sullivan's take on crab cakes, Maryland style. He is the executive chef at Harbor Magic Hotels.

1 tbsp. butter
3 tbsp. chopped scallions
2 tbsp. chopped parsley
2 oz. mayonnaise
1 tbsp. Dijon mustard
1 egg
1 tbsp. Old Bay seasoning
1 tbsp. Worcestershire sauce
Juice of ½ lemon
12 oz. crabmeat, back fin lump
4 oz. claw meat
2 oz. Japanese panko

Melt the butter in a small sauté pan and add scallions and parsley. Cook for 2 minutes and place in a small mixing bowl. Add the mayonnaise, Dijon mustard, egg, Old Bay seasoning, Worcestershire sauce, and the lemon juice and blend well using a whisk. In a separate bowl, combine the crabmeat and the breadcrumbs and carefully mix to ensure no breaking of the crab lumps. Add the sauce mixture and blend. Shape into 4 crab cakes and place on an oiled baking sheet. Bake at 350 degrees until slightly golden brown.

*(Courtesy Harbor Magic)*

# Chesapeake Bay Blue Crab Cakes

This recipe comes from Mike Hutt, the Virginia Marine Products Board's executive director. They use soft breadcrumbs and add the zest of a fresh lemon, cayenne pepper, and horseradish on top of the Old Bay seasoning.

⅓ cup fresh-made mayonnaise
1½ tbsp. Dijon mustard
2-3 large eggs
2 tbsp. lemon juice, freshly squeezed
1 tsp. seafood seasoning, such as Old Bay
½ cup fresh, soft breadcrumbs
2 tbsp. fresh parsley, chopped
2 tbsp. green onion, finely chopped
1 tsp. zest of fresh lemon
1 pinch of cayenne pepper
2 tsp. prepared horseradish
Salt and black pepper, to taste
2 lb. fresh Virginia jumbo lump blue crabmeat
Flour
Clarified butter or peanut oil for sautéing

Combine first 11 ingredients. Taste and adjust seasoning. Gently fold into crabmeat. Portion into 3½ oz. cakes. Lightly flour, sauté in clarified butter or peanut oil. Serve immediately.

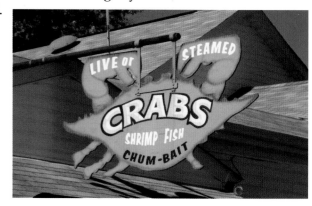

*Crab sign*

# Legal Sea Foods Crab Cakes

Heather Freeman, who handles the media and public relations for Legal Sea Foods, managed to persuade the chefs there to share their crab cake recipe that was voted reader's choice in the *Washingtonian* magazine in 2009.

Yields 2 crab cakes

4 oz. Ritz crackers
2 tbsp. butter
⅓ cup scallions, white and green parts, chopped
½ cup celery, diced
3 tsp. garlic, minced
1-2 tbsp. dry white wine
⅛ tbsp. Old Bay seasoning
1 tbsp. fresh lemon juice
1-2 tbsp. fresh parsley, chopped
8 oz. crabmeat, back fin
Salt and freshly milled black pepper
Chopped red peppers
1 large egg
Flour for coating/dredging

Place the crackers in a plastic bag and crumble with your hands or with a rolling pin. Put them in a large mixing bowl and set aside.

Melt butter in a large skillet over medium heat and cook the scallions, celery, and garlic for 5 minutes or until cooked through but still slightly translucent. Toss with the crackers.

Stir in the wine, Old Bay seasoning, lemon juice, parsley, and crabmeat. Season with salt and black and red peppers to taste. Mix in the egg. Refrigerate for at least 1 hour.

Once the mix has set, remove from refrigerator and divide mixture into 4 equal portions.

Form crab cake by gently forming with your hands into a loose "ball." Do not overpack; it should remain loose.

Dust with flour and sauté in olive oil or butter until cooked through.

# Roy's Baltimore Crab Cake Recipe

Roy Yamaguchi, the James Beard Award-winning chef, opened his first Roy's restaurant in Hawaii in 1988, creating Hawaiian fusion cuisine that combined European techniques and Asian cuisine. Fortunately, one of his thirty-one restaurants is here in Baltimore, where Patrick Crooks, chef/partner, satisfies your taste buds with Roy's Baltimore crab cake. Just leave room for the Melting Hot Chocolate Soufflé with raspberry coulis and vanilla bean ice cream.

If you don't have a ring mold, or muffin cutter, then cut both ends off a small tuna fish can or similar size can and use it as a mold. Just be careful when you run your finger around the inside of the can/mold. As the crab cakes are cooked in the ring mold, it would be ideal to have eight of them so you can cook them all at once. If you don't have eight, then prepare, cook, and keep warm until they are all cooked.

## Crab Cakes

1½ heaping tbsp. mayonnaise
1½ tbsp. lemon juice
2 egg yolks
1 tbsp. Worcestershire sauce
½ oz. Sriracha hot sauce
1 tbsp. parsley, finely chopped
1 tbsp. cilantro, finely chopped
1 tbsp. chives, finely chopped
Salt and pepper, to taste
1 lb. jumbo lump crab, pick without breaking up
Japanese panko, add enough to soak up the liquids in mix
   (firm, not dry)

## Assembly

Vegetable spray
2-inch ring molds or biscuit cutters
Parchment paper cut to fit the mold

Crab Cakes

Mix all ingredients except the crabmeat and panko. Gently add the crabmeat without breaking up and then the panko.

Assembly

Spray each individual ring mold with vegetable spray, running your finger around the inside to make sure that you get everything, then line the inside of the ring mold with the parchment paper and coat with vegetable spray.

Next fill your molds with crab filling, making sure it is leveled and packed (portion depends on your preference). Do not pack it so hard that the juices come oozing out the sides.

Place the mold or molds on a baking sheet and cook on the top shelf of broiler for 2 to 3 minutes or until brown. Next, using a spatula, remove from the broiler and carefully flip over to brown the other side. When brown, gently push the crab cake out of the mold and parchment and immediately place onto warm plate.

Serve with desired sauce or lemon wedges.

*Chef Roy Yamaguchi (Courtesy Roy's Restaurant)*

# Senator Barb's (Mikulski) Favorite Crab Cakes

As Senator Barbara Mikulski is now the longest-serving senator in the United States, her take on crab cakes is important.

Yields 4 crab cakes

1 egg, or egg substitute for special diets
2 slices white bread
1 tbsp. mayonnaise, light or regular
1 tbsp. Dijon mustard
2 tsp. Old Bay or Wye River seasonings
1 tbsp. parsley, snipped (optional)
1 lb. crabmeat, jumbo lump or back fin
Tartar sauce or cocktail sauce

Beat the egg in a bowl. Trim the crusts from the bread and tear the slices into small pieces. Add these pieces to the egg. Mix in the mayonnaise, Dijon mustard, Chesapeake seasoning, and parsley and beat well.

Place the crabmeat in a bowl and pour the egg mixture over the top. Gently toss or fold the ingredients together, taking care not to break up the lumps of crabmeat.

Form the cakes by hand into patties about 3 inches around and ¾-inch thick. Shape should be like a cookie, not like a meatball or golf ball. Place the cakes in the refrigerator for at least 45 minutes before cooking. This is very important so the cakes don't fall apart.

### Broil

Slip the crab cakes under a preheated broiler until nicely browned, turning to cook evenly, about 4 to 5 minutes on each side.

### Sauté

Heat a small amount of butter or olive oil in a skillet and sauté the cakes, turning several times, until golden brown or about 8 minutes total cooking time.

Serve at once with tartar sauce, mustard, or cocktail sauce on the side.

# Gibson Island Crab Cakes

Nancy Vaughan is a publicist of extraordinary talent who lives in the Phoenix area now but grew up in Northern Virginia. She raves about this recipe from her friend Pamela Freytag. Pamela says, "These will spoil you for any other crab cakes. They were a trade secret of the chef at the Gibson Island Club (near Annapolis). Without the club's knowledge, he entered his crab cakes in the First Annual Maryland Seafood Bake-Off. To his surprise (not to mention the club members), his crab cakes won first place! He had to divulge the recipe, which was then published, and he was fired. Club members were horrified, but they copied down the recipe even faster than those who'd never heard of Gibson Island.

The chef seems to remain anonymous if only to protect the guilty. Enjoy these guilty pleasures.

Yields 8 crab cakes

1 cup mayonnaise
1 egg
½ cup breadcrumbs
1 tsp. chopped parsley
1 tbsp. Old Bay seasoning
1 tsp. prepared mustard
Juice of ½ lemon
1 dash Worcestershire
2 lb. back fin crabmeat

Combine mayonnaise with all other ingredients except crab. Blend well. Fold in crabmeat. Form into 8 (5 oz.) balls and flatten each one slightly. Bake in a 500-degree oven (very hot) for 8 to 10 minutes until bubbly and golden (not brown).

*Live crabs sign*

# Crab Cakes and Polenta with Spicy Sweet Potatoes and Old Bay Crema

Adryon (pronounced Adrienne) says she is "currently confined to the suburbs but day dreams of a life on a big ol' farm." She has "tattoos, enjoy[s] caffeine, vodka, and burritos—although not simultaneously." From her Baltimore, Maryland, home she writes the Adryon's Kitchen blog and posted this recipe from her late husband who worked as a chef for a number of years. Among the things he taught her were "knife skills and the fun of inventing ways to use a new ingredient."

After traveling around as an army brat, she returned to Maryland because of the crabs and says she's a "sucker for a good crab cake. . . . What started as a craving for crabs and steamed corn spiraled out of control and thus was born the most wordy recipe title ever! Sorry about that. The combination of textures and flavors are out of this world."

**Tip:** Make the crema and crab cake mixture in advance and while the sweet potatoes are boiling, pop the crab cakes under the broiler and fry the polenta! I used premade (gasp!) polenta, so all I had to do was slice and toss it in the pan. Let's go!

Serves 4

## Crab Cakes

¼ cup mayonnaise
⅛ cup Dijon mustard
½ tsp. fresh lemon juice
½ tsp. hot sauce
1 tsp. Worcestershire sauce
1 tsp. Old Bay seasoning
1 tbsp. fresh chopped parsley
¼ tsp. garlic powder
¼ tsp. black pepper
¼ cup onion, minced
1 egg, beaten
½ cup white bread, torn to smithereens
1 lb. crabmeat

## Polenta

1 tube of store-bought polenta, or make your own
2 tbsp. olive oil
Black pepper

## Sweet Potatoes

3 large sweet potatoes, peeled and cut into uniform slices or cubes
¼ cup milk
1 tsp. chili powder
¼ tsp. cayenne pepper
Salt and pepper to taste

## Old Bay Crema

3 tbsp. mayonnaise
3 tbsp. crème fraîche or sour cream
2 tsp. Old Bay seasoning
1 tsp. Worcestershire sauce
1 tbsp. fresh chives, chopped
½ tsp. hot sauce
Black pepper to taste

## Crab Cakes

Combine all of the ingredients for the crab cakes except the crabmeat in a mixing bowl. Once all the ingredients are thoroughly combined, add the crabmeat and mix to combine. Allow the mixture to chill in the refrigerator for at least 15 minutes before using.

When ready to use, preheat your oven's broiler. Roll the mixture into 8 evenly sized crab cakes. Broil for 5 minutes on each side or until both sides are golden brown and hot throughout.

## Polenta

While the crab cakes are broiling, preheat 2 tbsp. of olive oil and toss 8 slices of store bought polenta (about ¼ inch) into the pan. Season with black pepper. Cook for 3-5 minutes on each side until golden brown. Remove from the pan and drain on paper towels to remove excess grease.

## Sweet Potatoes

Put the sliced sweet potatoes into a pot of cold water and bring to a boil. Salt the water and cook for 15 to 20 minutes, or until the sweet potatoes are fork tender. Drain and smash or whip the potatoes with the remaining ingredients. Taste and adjust seasonings. Set aside and keep warm until ready to use.

Place ½ cup of mashed sweet potatoes on a plate and top with a slice of polenta. Place a crab cake on top of the polenta and drizzle everything with the Old Bay Crema.

### Old Bay Crema

Combine all of the ingredients together in a bowl and chill for at least 30 minutes before serving to let all the flavors marinate.

*(Courtesy Adryon's Kitchen)*

## Nicholas Carroll's Spicy Crab Cakes

Nicholas Carroll is another Maryland ex-pat who misses his Maryland crab cakes. He's taken to extreme measures to approximate his memory for Asian style.

"It's really more a cooking method than a recipe, because I've tried it on my own Maryland crab cakes, a friend's, and (gasp) even Trader Joe's sorta Maryland crab cakes. It works on all of them.

"This was a desperation experiment the first time around. Lacking butter or any cooking oil except toasted sesame oil, I sautéed the crab cakes in that. I figured the taste would overwhelm the crab, so for the heck of it I added some Asian hot oil (red peppers steeped in oil). The result was sensational. The toasted sesame oil didn't overwhelm the crab, but complemented it. The hot oil didn't make the crab cakes spicy hot, just slightly spicy and comfortingly warm to the palate."

Nicholas says make your crab cakes as you wish and then:

Add enough toasted sesame oil to the frying pan to sauté (not deep-fry). Add 2 to 4 drops per crab cake of Asian hot oil to the sesame oil and mix thoroughly. Sauté as you would in butter until golden brown.

**Tips:** You "must" use toasted sesame oil. Plain sesame oil will not work. And you "must" use Asian hot oil. Mexican hot sauce will not work; it's tasteless until you add too much, and then it's far too spicy. Cayenne pepper in the toasted sesame oil might work if you let it meld for a week, but it is just harsh and hot if added right before cooking.

# Aunt Stell's Crab Cakes

Renee S. Gordon is a *Philadelphia Sun* travel writer whose husband Barry Gordon, says Renee, "fancies himself a gourmet chef. He is really a professor who cooks." Although he claims to be retired, Barry is an adjunct professor at Eastern University, Delaware County Community College, and Peirce College and consults for software applications training.

Barry says this recipe came from his Aunt Estella Beecham Vaughn who was born in 1888 and died in 1963. "Aunt Stell, as we affectionately called her, retired as a cook from the Atlantic Refining Company in Philadelphia, Pennsylvania. She was an excellent cook and a gracious lady. She was born in Pocomoke, on Maryland's Eastern Shore, and raised by Quakers in the Lancaster County area. One of her favorite sayings was 'Because you are small in stature doesn't mean you have to be small in mind as well.'" He notes that you may vary the amount of ingredients by your personal taste.

Yields 4 crab cakes

3 tbsp. olive oil
3 cloves garlic, diced
1½ stick of butter, divided
3 scallions, finely diced, white of all three and green of one
⅛ green or red pepper, finely diced
⅛ red onion, finely diced
3 tbsp. all-purpose flour
1 tsp. cornstarch
Crushed red pepper
Ground black pepper
Sea salt or regular salt
¼ cup white wine (Chablis)
1 tbsp. Old Bay seasoning
¼ cup plain breadcrumbs
1 lb. crabmeat, back fin
1 egg, beaten
¼ cup flour
1 tbsp. garlic powder
2 tbsp. extra-virgin olive oil
1 cast-iron frying pan (optional)

Add 3 tbsp. of olive oil to a pan under medium to low heat.

Sauté the diced garlic. Don't let the garlic burn; remove garlic when it begins to turn brown.

Add ¼ stick of butter to the pan and sauté the scallions, pepper, and onion until they are soft. Remove from the pan and then add 3 tbsp. flour and cornstarch to create a roux.

Return the sautéed vegetables to the roux. Add peppers and salt to taste. Add white wine and stir until slightly thick. Add Old Bay and then breadcrumbs and stir.

Gently add crabmeat and lightly toss all the ingredients together. Cook for 4 minutes on low heat. Let the ingredients stand and cool for 15 minutes.

Add beaten egg to bind the ingredients. Shape into crab cakes, any size you wish.

Mix ¼ cup flour, 1 tbsp. garlic powder, and salt, pepper, and Old Bay to taste. Dust each crab cake lightly with the mixture.

Refrigerate crab cakes for at lease 1 hour. Fry each crab cake until golden brown.

*Barry Gordon (Photograph by Renee Gordon)*

# Chesapeake Bay Crab Cake with Bloody Mary Salsa

Jimmy Schmidt, executive chef at Morgan's in the desert, the new signature restaurant at the legendary La Quinta Resort & Club in California's Coachella Valley, loves "traditional cooking methods, including open grilling, slow roasting, braising, pickling, and curing to create deliciously simple, rustic and healthful dishes."

The award-winning chef (possibly best-known for his Rattlesnake Club restaurants) is a "pioneer in America's culinary movement towards sustainable cooking and farm-to-table dining," and founded Chefs Collaborative in 1991, the nation's leading nonprofit chefs organization devoted to fostering a sustainable food system.

This recipe is a Westerner's take on a Maryland crab cake. It's quick and easy.

Yields 4 crab cakes

## Crab Cake

¼ cup mayonnaise
½ cup sweet onion, finely diced
¼ cup celery stalk, finely diced
Sea salt
Cayenne pepper, to taste
1 lb. Chesapeake Bay jumbo lump crabmeat, cleaned
½ cup whole-wheat crackers, ground
Parchment paper

## Salsa

1 cup ripe heirloom tomatoes, diced
1 tsp. garlic salt
1 shot Citron Vodka (optional)
1 lime juiced
1 tbsp. Lea & Perrins Worcestershire Sauce
1 generous dash Chipotle Tabasco, or to taste
1 heaping tbsp. prepared horseradish sauce
Sea salt, to taste
4 sprigs celery greens for garnish

## Crab Cake

Preheat oven to 375 degrees.

In a medium bowl combine the mayonnaise, onion, and celery. Season generously with salt and cayenne. Fold in the crabmeat. Mold into 4 cakes. Sprinkle half of the cracker crumbs into 4 small circles on a parchment-lined sheet pan. Place a crab cake on each circle of crumbs. Sprinkle the tops of the crab cakes lightly with the remaining crumbs. Deep chill in the freezer, without freezing, until ready to cook.

## Salsa

In another medium bowl combine the tomatoes, garlic salt, optional vodka, and lime juice. Season generously with Worcestershire sauce, Tabasco, horseradish, and sea salt as necessary. Reserve.

## Preparation

Place the crab cakes on the middle rack of your oven, cooking until golden brown for about 15 to 20 minutes. Remove from the oven. Transfer to the center of a warm serving plate. Spoon the salsa over the cake. Garnish with the sprig of celery and serve.

*Chef Jimmy Schmidt (Courtesy Morgans in the desert)*

*(Courtesy La Quinta Resort & Club)*

# Jumbo Lump Crab Cakes with Spicy Remoulade

Jeff Tunks is executive chef for DC Coast restaurant in Washington, D.C. Just as he and partners Gus DiMillo and David Wizenberg have preserved the Beaux-Arts building and made it a warm and exciting building that lives and breathes, you can almost feel the decorative influences in the cuisine. Tunks has lived and worked across the continent, from Georgetown to Coronado Bay to New Orleans, and Washingtonians are thrilled to have him here and contributing so much to the restaurant scene. Besides DC Coast, the guys own TenPenh, Ceiba, and Acadiana. As so many area residents aren't from here, they're sure to find some familiar cuisine at one of these restaurants.

Yields 5 appetizer-sized crab cakes

## Crab Cake

1 lb. jumbo lump Chesapeake Bay crabmeat
1 whole egg
1 tbsp. mayonnaise
Juice of ½ a lemon
1 tbsp. fresh chives, minced
Pinch cayenne pepper
Salt to taste
¼ to ½ cup fresh brioche crumbs (less is better)
3 oz. olive oil for sautéing

## Remoulade

4 tbsp. mayonnaise
1 tsp. horseradish
1 tsp. Creole mustard
½ tsp. Tabasco
1 tbsp. chili sauce
Juice of ½ a lemon
Salt and pepper to taste
½ tsp. Worcestershire sauce

## Crab Cake

Gently remove all shells from crabmeat, being careful not to break up lumps. In a large mixing bowl, combine egg, mayonnaise, lemon juice, minced chives, cayenne, and salt. Gently fold into cleaned crabmeat. Sparingly add brioche crumbs to lightly bind meat. Form into 5 cakes and reserve cold.

## Remoulade

Mix all ingredients together. Reserve.

## Preparation

Sauté crab cakes in olive oil until golden brown. Place in a 350-degree oven for 8 to 10 minutes. Place a crab cake in the middle of the plate. Top crab cake with a spoonful of remoulade.

*(Courtesy DC Coast restaurant, Chef Jeff Tunks)*

# Crab Cutlet (Crab Cake Variation)

This is a variation that includes a roux and the crab cake comes out with more of a meat cutlet texture than a crab cake.

¼ lb. butter
4 tbsp. flour
¾ pint cream
Dash dry mustard
Dash nutmeg
Dash salt
Dash pepper
1 lb. crabmeat
1 egg, beaten
½ cup breadcrumbs, your choice
Cooking oil

Make a white roux of the butter and flour, adding cream and seasonings. Allow to cool. Add the crabmeat and form into crab cakes. Put in a flat dish in the refrigerator to rest for at least 1 hour or overnight.

Dip the cakes into egg and roll in fresh breadcrumbs. Cook in oil until slightly golden, about 2 minutes on each side, and place on a baking sheet to heat through and brown in a 350-degree oven, about 10 minutes.

# SCK Maryland Style Crab Cakes

This crab cake recipe reflects Chip Ulbrich's Southern and Caribbean influences.

## Dressing
2 cups Duke's mayonnaise  
¼ cup Creole mustard  
2 tbsp. Old Bay seasoning  
Juice of 1 lemon  

## Crab Cakes
2 lb. jumbo lump crabmeat, picked clean  
¼ cup Italian parsley, chopped  
½ cup Japanese panko (more or less as needed), plus 8 cups for breading  
Salt and pepper, to taste  

## Dressing
Mix all ingredients together and set aside.

## Crab Cakes
Mix crab with parsley and enough dressing to moisten well. Add breadcrumbs (from the ½ cup portion) and try to form cakes. If they do not hold shape, add more crumbs or dressing until they do.

Form into 4-oz. cakes, then coat in reserved panko, pressing lightly to adhere.

Sauté in clarified butter on a griddle or in a nonstick pan until browned, flip, and cook on the other side until brown, about 3 minutes per side.

# Salads and Sandwiches

Crab salads and sandwiches allow you to use some leftover crabmeat (if there is such a thing) or to use types other than the back fin and lump crabmeat. The slightly stronger taste of the claw meat helps give the salad or sandwich a more flavorful tang. To some extent, whether you throw the mixture on a bed of lettuce or plop it onto some bread depends on how you feel and what you have in your pantry.

# Vintage Crab Salad

Mary L. Fugere, the director of media and community relations for the Hampton Convention & Visitors Bureau, says the Kecoughtan Literary Circle of Hampton has been in existence for more than one hundred years. The circle began in 1892 as an association of women who wished to study and keep abreast of the literary and scientific world. Except for the year 1917 when the membership devoted its time and dues to the war effort, the Circle has continued to meet and pursue its educational purpose.

Many of the current members have a familial connection to the original circle, particularly Mrs. James H. (Ann) Tormey whose great grandmother, Mrs. James S. Darling, was an early member. Mary Anna Daulman, was born in England in 1834 and came to New York in 1860. She met James Sands Darling, a ship joiner, and they were married in 1864. After the Civil War, business was poor so the Darlings, with two-year-old Frank, traveled to Hampton with a boat full of lumber. The city had been burned by Confederates and the Darlings planned to sell the lumber to people wanting to rebuild their homes. Unfortunately, no one had any money to buy lumber, so Darling explored other business interests and soon had a saw mill, a fish oil factory, and ultimately a successful oyster business. Darling's innovative business instincts would be key in the rebuilding of Hampton following the Civil War.

Mrs. Darling was highly respected for her business sense and domestic abilities and she left her "cook, scrap, and accounts book" to her great granddaughter, Ann. This ledger contained expenses; clippings of many recipes, including this crab salad recipe; directions for making socks and sheepskin rugs; and remedies for ailments.

Mary was kind enough to re-create the crab salad. She refined some of the measurements, so instead of a "small teacup full of vinegar" the recipe calls for ⅛ of a cup of vinegar. Cayenne pepper was in the recipe, and although she says it added color, Mary suggests replacing it with paprika or crab seasoning.

2 hard-cooked egg yolks
1 heaping tsp. sugar
1 tbsp. dry mustard
1 tbsp. sweet oil (mild vegetable oil)
⅛ cup vinegar
2 crackers, crumbled
1 tbsp. butter
½ tsp. celery seed*
Cayenne pepper or black pepper, to taste
Salt, to taste
12 oz. crabmeat

Blend the first 4 ingredients together until they become smooth. Gradually add the vinegar. Beat 2 crackers fine and mix in the butter, celery seed, and either cayenne or black pepper to suit the taste and a little salt. Blend in the crabmeat. Serve room temperature or cold.

*The celery seed, though small, must be beaten in a mortar or the pungent flavor will be lost.

*(Photograph by Mary L. Figere)*

# Basic Crab Salad

¼ cup mayonnaise
1 tsp. chopped pimiento
2 tsp. Dijon mustard
1½ tsp. Worcestershire sauce
½ tsp. salt
¼ tsp. hot pepper sauce
3 tbsp. lemon juice (about 1 lemon), retain lemon, sliced
⅓ cup chopped celery
1 lb. crabmeat
Lettuce
Dash paprika

Mix first 8 ingredients in a bowl. Add the crabmeat gently. Line bowl or platter with lettuce leaves, place crabmeat mixture in the center, sprinkle with paprika, and decorate with lemon slices.

# Open Gates Farm Bed & Breakfast Crab Salad

Jenna Licurgo, the innkeeper at Open Gates in rural Huntington, Maryland, says a friend gave her this recipe about fifteen years ago. He told her that when he was a boy growing up in Brooklyn, New York, he and his brothers would snack on salad rather than chips or popcorn while watching movies on TV. "I absolutely can't stand seafood, but this salad has such a mild crab taste that even I like it!"

½ cup mayonnaise
2 tbsp. white vinegar
2 tsp. dried oregano, crushed
1 small onion, diced
2 medium tomatoes, diced
1 cup crabmeat, diced
1 head of lettuce, iceberg or other, torn into pieces, or a large bowl of mixed greens

Combine all dressing ingredients and mix well. Pour over lettuce and mix thoroughly.

# Clyde's Crab Salad Tower

This tower recipe from the Clyde's Restaurant Group collection is time consuming and just might call for some ingredients that you don't have in your pantry. Among other things, you'll need eight four-inch ring molds (or cleaned tuna fish cans). The Pacific-influenced dish is well worth your time, though, and is certain to impress your friends.

### Crab Salad

1 lb. crabmeat, jumbo lump
1 lb. crabmeat, lump or back fin
1 cup mayonnaise
1 tsp. wasabi powder
2 tbsp. sweet-chili sauce
½ cup chopped chives
1 red bell pepper, finely diced
1 tsp. salt

### Coconut-Curry Rice Cake

2 cups short-grain sushi rice
1 14-oz. can coconut milk
3 cups water
3 tbsp. sugar
1 tbsp. salt
1 tbsp. green-curry paste
1 cup toasted, shredded coconut

### Citrus-Miso Vinaigrette

1 lemon, zested and juiced
1 lime, zested and juiced
2 oranges, zested and juiced
½ grapefruit, zested and juiced
3 shallots
2 garlic cloves
2 tbsp. fresh ginger, grated
1 cup white-miso paste
¼ cup honey
1 cup canola oil
1 tbsp. sesame oil
Salt, to taste

## Wonton Chips

8 wonton skins
Canola oil, to taste
Wasabi powder, to taste
Salt and pepper, to taste

## Crab Tower

Canola oil for frying
Goma wakame (sesame seaweed salad)
Crab Salad
Tobiko (flying-fish roe)
¼ avocado, fan sliced, for garnish
Sriracha hot sauce as needed, for garnish

## Crab Salad

Pick any cartilage or shell fragments from the crabmeat. Set aside.
In a large bowl, mix the remaining ingredients and then carefully fold in the crabmeat to avoid breaking up the lumps. Set aside.

## Coconut-Curry Rice Cake

Combine all the ingredients in a large pot set over medium-high heat. Cover, bring to a boil, then reduce to a slow simmer. Stir the rice occasionally with a rubber spatula to keep it from sticking to the bottom of the pot. Continue to simmer 25 to 30 minutes until the rice is fully cooked. On a sheet pan lined with parchment paper, spread the rice evenly, about 1 inch thick. Let fully cool in the refrigerator.

## Citrus-Miso Vinaigrette

Add all the ingredients except the canola and sesame oils and salt to a blender. While the blender is running at medium-high speed, slowly drizzle in the oils so the dressing emulsifies. Strain the dressing through a fine-mesh sieve and season with salt, if necessary. Set aside.

## Wonton Chips

Preheat oven to 300 degrees.
Cut the wonton skins diagonally. Mix the canola oil and wasabi powder and season to taste with salt and pepper. Brush the skins with the mixture. Lay the wontons on a sheet pan and bake until crisp, about 10 to 15 minutes. Set aside.

Crab Tower

Pour enough canola oil into a deep pot to cover the 1-inch-thick rice cakes. Heat the oil to 350 degrees.

Cut 8 4x4-inch pieces of parchment paper. On top of each piece, layer a ring mold with ¼-inch sesame seaweed salad, a thicker layer of jumbo lump crab salad (packed tightly), and a thin layer of flying-fish roe. Chill, preferably overnight, so they'll unmold easily.

From the tray of chilled sushi rice, use a ring mold to punch out 8 rice cakes. Deep-fry the rice cakes in the hot oil for 3 to 4 minutes—the outside should be crispy and the center soft and sticky. Set on paper towels to drain.

To serve the towers, spoon a small amount of the Citrus-Miso Vinaigrette in the center of 8 plates, then place a fried rice cake on top. Carefully remove the ring molds from the crab-and-seaweed patties (tilt them, remove the parchment, and slide the ring mold up while gently pushing the contents down with a spoon). Set on top of each rice cake. Top each crab tower with fanned slices of avocado. Gently insert a wonton chip into the avocado so it sticks straight up.

For garnish, add a few drops of Sriracha hot sauce around the plate.

*Crab table (Courtesy Shoreline Seafood, Gambrills, Maryland)*

# Hampton (Virginia) Crab Salad

This crab salad recipe comes from the Virginia Marine Products Board and takes almost no time to prepare.

1 lb. Virginia crabmeat
½ cup mayonnaise
2 tbsp. onion, finely chopped
1 cup celery, chopped
2 tbsp. olives, chopped
2 tsp. sweet pickle, chopped
2 hard-cooked eggs, chopped
½ tsp. salt
Dash pepper
Lettuce
2 tomatoes, cut in wedges
2 avocados, sliced

Remove any remaining pieces of shell or cartilage from crabmeat. Combine with mayonnaise, onion, celery, olives, pickle, eggs, salt, and pepper. Serve on a bed of lettuce with tomato wedges and avocado slices.

# Crab Spinach Salad

Fresh spinach combined with citrus, avocado, and crabmeat make a great spring or summer salad.

1 large avocado
1 tbsp. orange juice
1 lb. fresh spinach, cleaned and torn in pieces
1 cup crabmeat, lump
1 orange, peeled and sectioned (zest first for the salad dressing)
Salad dressing (see recipe below or use your own)

Prepare the salad dressing and chill.

Peel avocado and slice into rings, just before you're planning to assemble and serve the salad (so avocado doesn't turn, or prepare ahead of time and sprinkle with lemon, lime, or orange juice and wrap in plastic, leaving as little avocado exposed to the air as possible).

Combine the spinach, avocado, crab, and orange sections.

Toss gently with salad dressing.

## Salad Dressing

⅔ cup vegetable oil
⅓ cup orange juice
2 tbsp. sugar
½ tsp. orange zest
1 tbsp. vinegar
Salt, to taste
½ tsp. dry mustard

Combine the ingredients and shake.

*Steamed crabs*

# Crab Tacos

You have fast food tacos and homemade tacos and even taco salad, and you're happy. Now, it's time to try crab tacos with the crispy bite of fresh tomato and lettuce, a little melted cheese, and the sweetness of crabmeat. That's good casual eating. Of course, you could always break or cut the ingredients into bite size and toss as a salad instead of wrapping with your bread of choice.

1 tsp. olive oil
½ cup onion, chopped
8 oz. crabmeat
Juice of 1 lime (2-3 tbsp.)
8 taco shells or tortillas, or tostada shells
1 cup Jack or Cheddar cheese, shredded
1 cup lettuce, shredded
4 tomatoes, chopped
1 avocado, thinly sliced
Kosher salt and freshly ground black pepper
¼ green onion or scallion, thinly sliced
Cilantro, finely chopped

Preheat oven to 350 degrees.

In a large skillet, heat oil over medium heat. Sauté the onion until soft, about 4 minutes. Add the crab and warm for about 2 minutes (depending on the size of the crabmeat bits, cook a little longer for large back fin lumps and less for special or mixed). Add the lime juice.

Put taco shells or tortillas on a baking pan and warm in oven, about 3 minutes. Arrange on a platter and fill with crab mixture, dividing evenly. Top with cheese (so the warm crab has a chance to start melting it), lettuce, tomato, and avocado. Sprinkle tacos with green onion and cilantro. Serve immediately.

# Crab Tacos with Old Bay Slaw and Charred Corn and Avocado Salsa

As mentioned in the crab cake section, Adryon writes the Adryon's Kitchen blog and before she sent the crab cake recipe, she created this one with tacos.

She says, "My kiggity kid loves tacos. It was a long, tedious battle with her, but the glory of tacos finally won. As a baby she ate beans, onions, homemade gorditas, enchiladas, and anything you put in front of her. Over time her love of Tex-Mex dwindled to cheese quesadillas and not much else. On a whim, she tried a fish taco, and ever since then we've been back in the Tex-Mex game.

"Using crab meat can be just as versatile as any other seafood, as the flavor is mild and welcomes a world of flavor combinations. I think next time I'm going to try one with a more Asian flare—a ginger/soy slaw with a wasabi crema.

"The Old Bay slaw adds a cold crunch to the warm crabmeat, and the charred corn salsa is the matriarch of the taco family, bringing everyone together to party. This makes enough for 8 people to have 2 tacos. That's a lot—so feel free to halve the recipe!"

Serves 8

### Charred Corn and Avocado Salsa
3 cups of corn
2 avocados, peeled and diced roughly
1 medium tomato, diced
½ cup red onion, diced
1 jalapeño, minced
½ cup cilantro, chopped
Juice of 2 limes

Old Bay Slaw

6 tbsp. mayonnaise
6 tbsp. sour cream
3 tbsp. milk or cream
4 tsp. Old Bay seasoning
¼ tsp. black pepper
4 cups cabbage, shredded

Crab Tacos

2 lb. crabmeat, picked through to remove shells
18 soft tortillas, flour or corn

Charred Corn and Avocado Salsa

Set oven to broiler. Lightly spray a baking sheet with cooking spray/olive oil. Spread the corn on the baking sheet and place under broiler until the corn is browned but not burned, about 8 minutes.

Combine the corn with all the remaining ingredients and set aside so the flavors can mingle while you prepare the rest.

Old Bay Slaw

Combine the mayonnaise, sour cream, milk, Old Bay, and black pepper in a large mixing bowl. Add the cabbage and toss to combine. Set aside and chill until ready to serve.

Crab Tacos

In a large skillet over medium heat, warm the crabmeat through.

Place the warm crabmeat on a tortilla and top with Old Bay Slaw and the Charred Corn and Avocado Salsa.

*(Courtesy Adryon's Kitchen)*

# Founding Farmers Mini Crab Roll

As mentioned in the appetizer chapter, Chef Al Nappo is at the helm of Founding Farmers, said to be D.C.'s greenest restaurant. You can read more about him in that section. This is their take on a crab roll.

2 artisan hot dog buns
2 tbsp. unsalted butter, melted
5 oz. crabmeat, lump
4 tbsp. Founding Farmers Louie Dressing (see recipe)
4 tsp. celery, finely diced
4 tsp. jicama, peeled and finely diced
2 tbsp. thinly sliced green onion, whites only
2 pinch Old Bay seasoning

Shave off the ends and sides of hot dog buns to make each into an even rectangular shape. Make a ½ inch deep cut through the middle of each bun, creating a "channel." Evenly brush the sides of buns with melted butter and toast on a hot griddle pan.

While the buns are toasting, mix the crabmeat with the Louie dressing, celery, jicama, and green onions. Remove the buns from the griddle pan and place on a plate. Divide the crabmeat mix in half and place in each "channel" of the buns, allowing the mix to fall over the sides. Garnish each with a pinch of Old Bay seasoning and serve immediately.

*Neon crab sign*

# Crab Po' Boys

This is a Chesapeake Bay version on the traditional New Orleans staple.

½ cup onion, minced
½ cup red bell pepper, minced
4 tbsp. unsalted butter
⅔ cup mayonnaise
2 tbsp. sweet pickle, finely chopped
1 tbsp. Dijon mustard
½ cup scallion greens and bulb, finely minced
1 lightly beaten egg white
1 cup fine, dry breadcrumbs, divided
1 tsp. Worcestershire sauce
1 lb. crabmeat, lump
Crab seasoning, to taste
Salt, to taste
4 rolls split-top bread, lightly toasted
Lettuce

Sauté the onion and pepper in the butter until they're soft. Combine the onion/pepper mixture with all but the last four ingredients, using half of the breadcrumbs and saving the crabmeat until last to gently fold into the rest of the mixture. Add crab seasoning and salt to taste.

Split the bread and toast lightly, if you wish. Line with lettuce and crab mixture.

*Clyde's Restaurant Group Maryland Crab Soup (Courtesy Clyde's Restaurant)*

# Soups

Crab soups are great all year, served cold in the summer and warm in the fall and winter. Because a lot of people don't bother eating the little legs and swimming legs, save them to make a stock. Throw them in a pot of water (six to eight quarts, depending on how many crab legs you have and how much stock you want to make), add a large chopped onion, a couple of carrots, a few stalks of celery (yes, include the leaves), peeled garlic cloves, and some salt and pepper. Simmer for about two hours and skim off any foam that may rise to the top. Strain the broth-based or clear (not cream) soup and chill. Pour into ice cube trays and freeze, and then put the cubes into large plastic freezer bags so you have crab stock whenever you need it.

Up until a few years ago when Thai and other international ingredients became fairly readily available in the U.S., crab soups came in two types. One had a cream base (New England style) and the other a tomato base (Manhattan style). Purists will claim there is only one way to make crab soup. Fortunately, there are many options.

She-crab soup does not refer to the gender of the crab being cooked. (Unless you caught them or brought them home to clean them, how on earth would you know?) According to Randy Howat, whose late father, Mark Howat, was a restaurant reviewer for the *New Jersey Record* for forty years, "She-crab soup is called 'she' because originally the recipe called for roe to flavor and color the soup."

It should go without saying, but I'll say it again. Gently check the crabmeat for cartilage before adding it to your soup.

# Clyde's Restaurant Group Maryland Crab Soup

The late Stuart Davidson, a Harvard graduate, World War II pilot, and international businessman, opened the first Clyde's in Georgetown in 1963. There are now thirteen restaurants in the D.C. area (including Northern Virginia and Maryland). The Columbia, Maryland, restaurant opened in 1975, and, yes, that could be movie mogul Barry Levinson sitting there enjoying lunch. While the restaurants are different, the recipes are consistent. This is their version of Maryland crab soup, modified to be made at home. There are a lot of ingredients and some prep time involved, but only a few steps and the cooking is done in thirty minutes. *makes ~12 servings*

- 1 lb. carrots, small diced
- 1 lb. onions, small diced
- 1 lb. celery, small diced
- ½ lb. butter
- ½ gallon chicken stock
- 23 oz. clam juice
- 3 tbsp. Old Bay seasoning *[4]*
- ¼ cup Worcestershire sauce
- ½ tbsp. black pepper
- 1½ qt. small diced potatoes — *takes a while to cook through*
- 3 lb. canned diced tomatoes in juice
- 1 tbsp. dry mustard
- ¼ cup horseradish
- 13 oz. tomato juice
- ½ lb. diced fresh green beans — *cook a little*
- ½ lb. corn, fresh or frozen
- 1 lb. crabmeat, lump or claw

In a large pot, sauté the carrots, onions, celery, and in butter for 10 minutes. Add the remaining ingredients except for the beans, corn, and crabmeat. Simmer for 20 minutes or until the potatoes are done Add the beans and corn. Add the crabmeat to the soup and heat through, trying not to break up the crabmeat too much. Serve in a large soup bowl with oyster crackers or crusty baguette.

# Cold Crab and Avocado Soup

I spent a little time working at the National Academy of Sciences in Washington, D.C., where brilliant minds are brought together to study today's interesting topics. Part of my job was coordinating the menu for the periodic face-to-face meetings, and during the summer I loved serving a cold soup to start the meal. This one is colorful, takes almost no time to fix, and should be made the day before, leaving you with time for other things on the day it will be served.

Serves 6

2 ripe avocados, peeled and pitted
½ cup chicken stock
½ tsp. ginger
1 cup half-and-half, or milk
Salt, to taste
Pinch of cayenne
6 oz. crabmeat, lump
Thin slices of lemon

Cut avocados in chunks. Place all ingredients except the crabmeat and lemon slices into a blender or food processor and blend until smooth. Chill at least 1 hour, overnight is better.

Place lump of crabmeat into each bowl, cover with cold soup and add a lemon slice. Adding mango to the recipe adds an interesting sweetness to the soup.

# Golden Crab Bisque

Jerry Edwards, chef at Chef's Expressions, self-billed as a company that creates "fashionable catered events," created this bisque in 1986. He developed it for a Baltimore's best crab soup competition, and as he's not from Baltimore, he figured he didn't have much of a chance. Then, as he says, "A customer came by to tell me she had entered the competition herself, but they were not allowing her to compete since she was not a professional chef. As a favor, I stepped in to take her place. The night before the competition, she came by my kitchen to try and help me develop a winning crab soup recipe that did not contain the ingredients that most crab soups included like Old Bay seasoning and potatoes. I remember her disdain when I added garlic to the pot. I told her of the garlic crabs I would make with my mother when I lived north in Philadelphia. We would toss fresh crabs, uncooked, in hot olive oil, garlic and red chili pepper flakes. After her first taste of the bisque, she proclaimed 'that's it, it's a winner.'"

To his surprise, "the combination of garlic-perfumed crabmeat and creamy rich havarti laced cream put us at the top of the competition that day. We celebrated our victory and made this bisque a staple in our kitchen ever since. We have never changed this recipe; just the presentation."

Edwards has won a slew of local and national awards for his talents and was recognized as an outstanding young (when he was still under thirty-five) chef of Baltimore. He says he prefers serving the soup between July and October when Bay crabs are sweet, but he notes with crabmeat available all year, you should "try this by the fire with a glass of fine sherry or a crisp Sauvignon Blanc."

*Twenty-fifth anniversary crab bisque (Courtesy Chef's Expressions)*

Serves 12

1 oz. butter, softened
¼ cup flour
2 oz. butter, chilled
2 ribs celery, minced
½ yellow onion, minced
4 cloves garlic, smashed and minced
4 tsp. sea salt
1 tsp. freshly ground pepper
6 oz. dry sherry
2 qt. heavy cream
10 oz. havarti cheese, thinly sliced
Juice of 2 lemons
1½ lb. crabmeat, jumbo lump
¼ cup parsley, chopped very fine

Make beurre manié by working together the softened butter with the flour. Work into dough, and set aside.

Melt the 2 oz. of butter in a soup pot. Add celery, onion, and garlic. Sauté until vegetables are translucent. Add all of the salt and half of the pepper. Gently deglaze with ⅔ of cooking sherry. Add the heavy cream and stir well with a whisk. Cook to just short of a boil. Add the beurre manié and stir with a whisk until thick.

As the mixture again approaches a boil, add the cheese, lemon juice, remaining ounce of cooking sherry, and remaining pepper. When thickened to the consistency of rich cream, add the crabmeat, stirring but being careful not to break up the lumps. Cook over medium-high heat, stirring constantly until just before boiling. Soup should be hot, but should not boil. Top with parsley.

*Ocean Grille crab sign*

# Crab Bisque

Master Chef John Mason has been associated with Kurtz's Beach in Pasadena, Maryland, since he and his first cousin Bonnie took over the family business (opened in 1933) in the early 1990s. Other relatives—mothers, fathers, aunts, uncles, brothers, sisters, nephews, nieces, and others—work there. Yes, it's a family operation, and they hope you feel you're eating in your grandmother's kitchen when you dine at Kurtz's.

This is a dairy-based version of crab bisque that uses and takes advantage of meat from all parts of the crab (lump, claw, and special).

Serves 16

**Vegetable spray**
**2 sticks butter (½ lb.)**
**2 cups flour**
**½ gallon whole milk**
**1 gallon light cream**
**4 tbsp. Old Bay seasoning**
**⅛ tsp. cayenne pepper**
**1 tsp. salt**
**½ tsp. black or white pepper**
**2 lb. Maryland blue crabmeat**

Spray 8-qt. saucepan with vegetable spray. Make a roux with the butter and flour by bringing the butter to liquid over medium heat, then adding all of the flour at once, stirring continuously until smooth. The final product will become the color of light peanut butter.

Add milk and cream, raise the heat and bring to scorch, stirring the entire time. Lower the heat to medium; add all the ingredients except the crabmeat. Stir often and simmer for 30 minutes. Add crabmeat a little at a time. Stir until all crabmeat is added (takes a while). Simmer another 30 minutes, stir often during this time. Serve hot.

# Quick Crab Bisque

This is a variation on the previous bisque that you can add sparkle to with some fresh mushrooms and asparagus. It also starts with already prepared ingredients that you just whip up and blend together, and if you won't tell that you didn't start from scratch, then neither will I.

Serves 6

1 11.25-oz. can cream of mushroom soup
1 11.25-oz. can cream of asparagus soup
2 cups milk
1 cup half-and-half
1 cup crabmeat
¼ cup dry sherry

Combine the first 5 ingredients and heat just to the boiling stage. Add the sherry and serve.

*Crab bushels color-coded for date of arrival*

# Roasted Corn and Crab Chowder

This is the award-winning corn and crab chowder as prepared by Chef Gary Beach. He suggests making this a day ahead of time, heating and adding the crabmeat at the last minute. This will make a lot of soup, so be sure to invite everyone.

Yields 1 gallon

1 lb. melted butter
1 lb. flour

Stock

2 qt. aseptic, low-sodium chicken stock, or 2 qt. crab base or crab stock (ask your local restaurateur for crab base, follow directions)
2 bay leaves
2 dashes hot sauce
2 dashes Worcestershire sauce
¼ cup celery, minced
¼ cup carrots, diced
¼ cup red peppers, diced
½ cup sweet onion or shallots, diced
½ tbsp. garlic, minced
Dash fresh lemon juice
1 cup white wine
1 dash ground white pepper
1 dash ground red pepper
2 dashes Old Bay seasoning
½ dash ground nutmeg
1 pinch ground black pepper

2 qt. heavy cream
2 lb. potatoes, peeled and cubed
6 ears roasted corn off the cob, or 2 15-oz. cans corn kernels, drained
1 or 2 lb. crabmeat (I like claw meat or ½ back fin and ½ claw)

Make a roux by heating together slowly equal parts butter and flour until smooth. Set aside to cool.

Add all stock ingredients to a stockpot. Boil wildly, then reduce to simmer when veggies are soft.

Slowly whisk in 2 qt. heavy cream. Add potatoes and corn. Cook on medium high while stirring, about 10 minutes, then reduce to a simmer.

Add the cooled roux slowly, making sure not to over thicken (if you do, add more chicken stock, milk, wine, half-and-half, water, or heavy cream or a combination). Simmer on low. Add crabmeat right before serving (crabmeat will burn and stick to the bottom otherwise). Adjust seasoning to taste (may need to add Old Bay seasoning, J.O. seasoning, and/or salt and pepper).

# Cold Crab Claw Soup

Here's another cold soup, perfect for hot summer nights, which can and probably should be prepared the day before serving. It takes advantage of the slightly stronger taste and less expensive price of the claw crabmeat.

1 lb. crab claw meat
2 tbsp. butter
3 tbsp. chopped celery leaves
½ cup sliced scallions, green parts
½ tsp. Old Bay seasoning
1-2 shakes Worcestershire sauce
2 cups half-and-half
1 cup milk
Salt and pepper to taste

Slightly sauté the crabmeat, celery leaves, and scallions in butter for 2 to 3 minutes. Add the Old Bay, Worcestershire sauce, half-and-half, and milk. Stir until hot and add salt and pepper to taste. Do not let it come to a boil. Let stand for 1 hour or longer. Serve hot or cold.

You can substitute the milk and cream with chicken, fish, or vegetable stock for an even lighter taste.

*Seafood store*

# Pumpkin Crab Soup

Pumpkin season and crab season overlap, so this is a great way to transition from summer fun to brisk fall days calling for hearty soups. Use your favorite pumpkin or a can of pumpkin purée (not pie filling). Once you feel comfortable making pumpkin soup (and with canned pumpkin and canned crab available all year, you can do this whenever there's a chill in the air), you can modify your ingredients.

2 lb. pumpkin, or about 2 cans of pumpkin purée
3 tbsp. olive oil
1 large onion, cut in ¼ inch slices
1 tsp. nutmeg
1 tsp. paprika or cayenne pepper
1 tsp. basil or dill
½ tsp. salt
6 cups chicken stock, homemade or store bought
¾ lb. crabmeat, lump, special, or a mixture
4 oz. heavy cream, half-and-half, or milk

If using fresh pumpkin, cut and bake at 350 degrees for 45 minutes and then purée. Set aside.

Heat the oil in a skillet and add the sliced onions. Sauté for 5 to 8 minutes. Add the nutmeg, paprika or cayenne, basil or dill, and salt (if your stock is low-sodium; otherwise omit the salt). Stir for 1 minute. Add the chicken stock and pumpkin. Simmer covered for 15 to 20 minutes. Purée the soup in a blender (you may have to do this in batches). Return soup to pot. Add the crabmeat and simmer for another 3 to 5 minutes. Add cream and stir.

For variations, use a medium butternut, acorn, or spaghetti squash; leeks; or brown sugar. Instead of sautéing the onions and leeks, try roasting them with the pumpkin/squash. Another option is to use coconut milk instead of heavy cream, cilantro, and maybe some Chinese Five Spice for an Asian flavor. Add some chopped and sautéed celery and other fall vegetables for an earthier taste. For another last-taste-of-summer-any-time-of-the-year option, add about ½ cup of corn, either fresh off the cob, frozen, or from the can. Bacon, everyone's favorite food group these days, can be thrown into the mix. Start with maple syrup or apple flavor and then venture into pancetta, hickory or corncob smoked, and particularly cinnamon.

# 4th & Swift's Summer Sweet Corn Soup with Lump Crab, Chives, and Old Bay

Jay Swift, the chef and owner of Atlanta's 4th & Swift, has been in the culinary business for more than thirty years. He's originally from Baltimore and graduated first in his class from the American Culinary Federation Apprenticeship program. In 2009, *Atlanta* magazine named his restaurant one of the top fourteen best news restaurants in the city.

10 ears of sweet yellow corn
1 Vidalia onion
3 tsp. vegetable oil
1 qt. vegetable or chicken stock
1 cup fresh heavy cream
Salt and white pepper, to taste
1 oz. steamed lump or jumbo blue crabmeat per serving
1 tbsp. fresh chives, sliced
4 oz. crème fraîche, or sour cream
1 tsp. Old Bay seasoning

Remove corn and juice from the cob and discard the cob. Peel and chop the onion. Bring a large saucepan or frying pan to low to medium heat and add oil. Sauté the corn and onions while stirring but do not allow to brown. Cook for 4 minutes.

Add half the stock and bring to a simmer. Turn off and let it cool down a bit.

With a ladle, scoop most of the corn into a high-speed blender and add more stock if too thick. Fill blender to below fill line, approximately halfway (this may require doing twice depending on size of blender). Add fresh, heavy cream. Place the top back on the blender and cover with a towel. Carefully pulse on low at first to get started safely, then blend at high speed until velvety smooth.

Strain soup and return to stove. Bring to a simmer and serve. Garnish with jumbo lump Crab, fresh cut chives, a dollop of crème fraiche, and a sprinkle of Old Bay seasoning.

# Crab and Tomato Soup

This is a delightful summer soup that takes advantage of the excess tomatoes growing in your garden. If you're buying tomatoes, you can try a combination of beefsteak, plum, and Romas or whatever the grocer recommends as a good soup tomato.

Serves 8

1 qt. chicken, fish, or crab stock
3 lb. fresh tomatoes, quartered, or use canned tomatoes
1 cup corn, fresh, canned, or 8 oz. package of frozen corn, thawed
1 cup potatoes, diced
¾ cup celery, chopped
¾ cup onion, diced
¾ tbsp. seafood seasoning
1 tsp. salt, adjust depending on the sodium content of your stock
¼ tsp. lemon pepper
1 lb. crabmeat, regular or claw

Place stock, vegetables, and seasonings into a large pot and simmer, covered, for about 45 minutes or until vegetables are almost done. Add the crabmeat, cover, and simmer for 10 more minutes. Serve hot or refrigerate for a day and serve as a cold summer soup.

*Crab trap and gear*

# Cream of Crab Soup

Thomas "Tom" M. Meyer is in charge of developing new projects and refining existing restaurants for Clyde's Restaurant Group. He became corporate chef for the company in 1983 and has spent a lot of time "fine-tuning the great American saloon menu. Dedicated to cooking in season, he has been passionate in his pursuit of exceptional ingredients, knowing that what comes in the back door is where it all begins." After earning a degree in hotel and restaurant management from the State University of New York, he decided that the action and fast pace of working in a kitchen was much more fun and satisfying than managing a hotel. He then attended the Culinary Institute of America, graduating in 1981.

"All around the Chesapeake Bay, cream of crab soup is on the menu," says Meyer. "All the small restaurants and the crab shacks serve it. The soup is simple and tasteful, always seasoned with ubiquitous Old Bay seasoning. It features the sweet succulent meat from the blue crab. Like for chowder along the coast of New England, there are as many recipes as there are cooks for cream of crab."

This is his take on cream of crab soup. He says the soup improves upon standing. When reheating, heat slowly and do not bring to a boil. Two or three ounces of sherry are a nice addition to this soup.

Serves 6 to 8

4 tbsp. butter
½ cup finely minced onion
½ cup flour
1½ cups chicken broth
1 cup heavy cream
1 cup milk
½ tsp. Old Bay seasoning
½ tsp. salt
¼ tsp. white pepper
¼ tsp. celery seeds
1 lb. crabmeat, lump, cartilage removed

Melt the butter in the top of a double boiler. Add the onions and cook until they are soft. Add the flour and stir with a wooden spoon until combined. Add the chicken broth; stir frequently until the mixture thickens. Add the cream and milk. Keep stirring. The soup should thicken enough to coat the back of a teaspoon. This will take at least 20 minutes. Add the Old Bay, salt, pepper, and celery seeds. Stir in the crabmeat.

*Jumbo crabs for sale*

# Easy Peasy Cream of Crab Soup

Vanessa Gill is the director of marketing for the marvelous Calvert Marine Museum in Solomons, Maryland. When she sent this recipe, she noted, "I had a traditional steamship recipe from an Annapolis grandmother. It was much more involved and took me at least four hours to make. My girlfriend from the Eastern Shore tasted it and said it was good, but not nearly as good as hers. So I challenged her to beat me! And she did . . . with this recipe, and it only took 20 minutes for her to kick my butt! Now I bring the soup to our family holidays—Easter, Thanksgiving, and Christmas—and I'm never asked to make anything other than that. They think I slave over it for days and wonder how I get the cream so perfect. . . . I've been told I should sell my recipe to local restaurants! I'm sure Campbell's Soup wouldn't appreciate that very much!"

1 onion, diced small
2 tbsp. butter
1 can cream of shrimp soup
1 can cream of celery soup
16 oz. heavy whipping cream
1 tbsp. Old Bay seasoning
¼ cup sherry
½ lb. crab

Sauté the diced onion in butter until soft in a soup pot. Add soups and cream to the onion. Bring to a low simmer. Add Old Bay and sherry to taste. Add crab. Serve hot. The entire soup takes about 10 minutes, but tastes like it took all day!

# She-Crab Soup

Chef Shannon "Red" Overmiller is a Patuxent River and Bay area native who creates a majestic she-crab soup at her Majestic restaurant in Old Town Alexandria, Virginia. She attended Maryland's esteemed L'Academie de Cuisine and worked at the Restorante Tosca and Restaurant Eve, and then spent time in Italy where she mastered her pasta techniques. All along she developed a locavore's passion for fresh ingredients. According to Overmiller, "This soup is even better the next day."

Serves 4 to 6

½ lb. butter, divided
1 cup onion, diced small
½ cup celery, diced small
Salt and freshly ground pepper, to taste
½ cup flour
2 tbsp. Worcestershire sauce
1 tbsp. Tabasco sauce
½ cup sherry
2 qt. half-and-half
1 lb. Maryland jumbo lump, blue crabmeat, shelled
Hard-cooked egg yolk, garnish

In large pot, heat ¼ lb. butter along with the onion and celery. Add salt and pepper to taste. Sweat until translucent and tender.

Add flour, then cook until blond in color and a smooth consistency is achieved. Add Worcestershire and Tabasco. Add sherry, then reduce. Once reduced by about one-third add half-and-half. Add crabmeat and heat to just below boiling, but do not boil. Add remaining butter at the end to enrich. Check seasoning. Crumble a little hard-cooked egg yolk on top of the soup as imitation roe and to provide a little color. Serve and enjoy.

*Chef Sharon Overmiller, Majestic restaurant (Photograph by Meshelle Armstrong)*

# South City Kitchen Old Charleston She-Crab Soup

Chip Ulbrich is a chef who combines a Southern approach to food with his European-style culinary apprenticeship. He has worked at such establishments as the Woodstock Inn & Resort in Vermont; Little Dix Bay Resort in Caneel Bay on St. John, U.S. Virgin Islands; and more. That's an award-winning background and he's been generous enough to send along this recipe with a Charleston, South Carolina touch.

Serves 4 to 5

½ lb. unsalted butter
1 medium-sized yellow onion, diced
4 ribs celery, sliced
1 tbsp. garlic, chopped
2 bay leaves
1 tsp. fresh thyme, chopped
½ tsp. nutmeg
3 tbsp. Old Bay seasoning
2 tbsp. Worcestershire sauce
2 tbsp. hot sauce
1 cup all-purpose flour
½ gallon whole milk
2 cups heavy cream
2 cups crab stock or clam juice
Salt and fresh black pepper, to taste
1 lb. blue crab claw meat, picked clean of shell
1 cup dry sherry

Melt the butter in a heavy-bottomed stockpot and sauté the onions, celery, and garlic for 5 minutes until softened. Add seasonings and sweat for a few more minutes.

Stir in the flour and whisk until smooth to create a roux, which will thicken the soup. Cook for 5 minutes over low heat, stirring frequently.

Add all other ingredients except for the crab and sherry. Whisk thoroughly together to remove all lumps. Let this come to a good simmer, stirring frequently to keep from scorching. Add more milk or stock if needed to adjust consistency; let simmer to cook flour in roux for about 10 more minutes. Strain through a fine sieve and return to heat.

Season to taste with salt and black pepper and finish by gently stirring in the crab and sherry.

Adjust seasoning and check consistency.

*(Courtesy South City Kitchen)*

# She-Crab Soup with Sherry and Whipping Cream

I'd guess you could find as many recipes for she-crab soup as you can find cooks who make it. Modifying a recipe to reflect your taste is the pleasure of cooking. Here's another version, without the chopped vegetables.

6 tbsp. butter
1 tbsp. all-purpose flour
2 cups light cream or half-and-half
1 cup milk
1 tsp. Worcestershire sauce
1 tsp. salt
1 tsp. grated lemon peel
1 lb. crabmeat
¼ cup dry sherry
½ cup whipping cream, whipped
Paprika, garnish
2 hard-cooked egg yolks, crumbled

Melt the butter and flour in the top of a double boiler to make a roux. Add the half-and-half and milk; then, while stirring, add the Worcestershire sauce, salt, lemon peel, and crabmeat. Cook slowly for 20 minutes. Season to taste.

Pour into soup bowls and add sherry. Top with a little whipped cream and paprika, and then crumble the yolk on top of the whipped cream.

# Lemon Crab Soup with Vegetables

This soup is neither tomato- nor cream-based, proving how much crabmeat has become part of everyone's cuisine.

1 pt. chicken, crab, or fish broth
1 clove garlic, peeled
1 cup mushrooms, sliced (your favorite type)
½ cup zucchini, thinly sliced
1 stalk (minus the bulb) lemongrass, cut into 4-inch sections
1 package fresh spinach, or 1 package frozen spinach, thawed and drained
8 oz. crabmeat
Zest of 1 lemon
1 tbsp. fresh lemon juice

In a medium saucepan, combine the broth and garlic. Cover and bring to a boil over high heat. Add the mushrooms, zucchini and the lemongrass. Reduce the heat and cover, cooking at a simmer for about 3 to 5 minutes.

At the last minute, add the spinach, crab, zest, and lemon juice. Remove from the heat.

If you have used low-sodium chicken broth, you may want to taste the soup and add salt and pepper. If you used regular broth, then the soup should not need additional salt and pepper. Remove the garlic and lemongrass.

# Crab Asparagus Soup

Fresh asparagus has its devotees. Even if you aren't one of them, try fresh white asparagus and you will click the "like" button on the asparagus fan page (if it has one). Add crabmeat and this is a perfect soup for a cool spring evening.

Serves 6

1 qt. chicken broth
3 egg yolks, beaten
1 bunch asparagus, cut into 2-inch pieces, or 1 14-oz. can of cut asparagus, drained
2 tbsp. soy sauce
3 tbsp. cornstarch
¼ cup water
8 oz. crabmeat

Heat the chicken broth and add ¼ cup to the beaten egg yolks. Slowly stir the yolks/broth mixture back into the broth. Add the asparagus and soy sauce. Combine the cornstarch and water to make a smooth paste. Add to the soup, stirring constantly until smooth and slightly thickened. Stir the crabmeat into the soup and heat to just before a boil. Lower the heat and simmer for about 5 minutes.

*Crabs today sign*

# Spicy Crab Soup

This super easy spicy crab soup comes from the Maryland Seafood Marketing and Aquaculture Development Program and is sure to warm you on those cold winter days.

Serves 8

1 qt. water
3 chicken parts, neck or wing
¾ cup onion, diced
3 lb. canned tomatoes, quartered
8 oz. frozen corn, thawed
1 cup frozen peas, thawed
1 cup potatoes, diced
¾ cup celery, chopped
¾ tbsp. seafood seasoning
1 tsp. salt
¼ tsp. lemon pepper
1 lb. Maryland crabmeat, regular or claw, fresh or pasteurized, cartilage removed

Place water and chicken in a 6 qt. soup pot. Cover and simmer over low heat for at least 1 hour. Add vegetables and seasonings and simmer, covered, over medium-low heat for about 45 minutes, or until vegetables are almost done. Add crabmeat, cover, and simmer for 15 more minutes or until hot. (If a milder soup is desired, decrease the amount of seafood seasoning to ¼ to ½ tsp.)

# Sauces

Some dishes demand or brighten when a sauce is added, so here are three, one from Marcus Bradley, the chef of Irish Eyes Pub in Lewes, Delaware; one from Kirk G. Plummer, the fish department manager at the Giant Food store in Gambrills, Maryland; and one from Elizabeth Fournier, from Cornerstone Funeral Services and Cremation in Boring, Oregon.

*Spicing crabs*

# Chesapeake Aioli

Marcus Bradley of Irish Eyes Pub, in Lewes, Delaware, makes this aioli for dipping. It's similar to a regular aioli with the addition of crab seasonings, either Old Bay, J.O., or your favorite spice blend. You could also (or instead) add the juice of one lemon and some lemon zest.

Marcus has spent a lot of time cooking in North Carolina, Washington, D.C., and New Orleans, so he's been around crabs for a long time.

½ cup mayonnaise
1 clove garlic, minced
1 tbsp. chives, minced
3 tbsp. of crab seasoning

Mix the ingredients in a bowl and refrigerate, covered, for at least 30 minutes or longer. Serve with crab claws, as a salad dressing, or whenever you need a dipping sauce.

# Crab Remoulade

A good fish department manager is a treasure, and Kirk G. Plummer, who works at the Giant Food store in Gambrills, Maryland, fits into that category. I mentioned to him that I was adding recipes to my collection, and he said he had a great recipe for a quick remoulade that is perfect for spreading on sandwiches. The secret is using mayonnaise (he prefers Hellmann's) so you don't have to worry about creating the egg and oil emulsion. It's already there and it won't break on you.

2 cups good-quality mayonnaise
1 tbsp. fresh parsley, chopped
1 small jalapeño, seeded and finely chopped
1 small shallot, thinly sliced
2 tbsp. sweet relish
2 tbsp. capers
2 tbsp. Dijon mustard
Juice of ½ a lemon
1 lb. crabmeat

Mix first 8 ingredients together and then fold in the crabmeat. Refrigerate for 30 minutes before serving. Use as a dip or a sandwich spread.

*Kirk G. Plummer*

# Green Onion Dipping Sauce

Elizabeth Fournier, from Cornerstone Funeral Services and Cremation in Boring, Oregon, occasionally serves food to her grieving clients. She says she adores using "Chesapeake Bay blue crabmeat in recipes ever since I ate them on a visit with cousins to Annapolis. This is a recipe I whip up when I know a Creole family will be coming to my country funeral home. I like to create dishes that suit the backgrounds of the visitors to my parlor."

As a West Coast version, Elizabeth boils her crabs with Creole seasonings—you can find her boil recipe in the introduction section—and then adds this sauce for a New Orleans' style lagniappe. I hope you try this recipe before you need Elizabeth's services.

3 bunches green onions
2 tbsp. minced garlic
1 qt. chicken stock
Salt, to taste
Freshly ground black pepper, to taste

In a saucepan, combine the green onions, garlic, and chicken stock. Season with salt and pepper. Bring the liquid to a simmer and cook for 10 minutes, then remove the sauce from the heat. Using a hand-held blender, purée the sauce until smooth. Strain the sauce through a fine mesh strainer. Reseason with salt and pepper if needed. Pile the crabs on a platter and serve with the dipping sauce. Scrumptious!!

# Entrées

Just because you've stowed the crab steamer and hammers and bibs for the winter doesn't mean you can't enjoy crab dishes. Although some recipes may require a trip to the grocery store, most of them will call for ingredients already in your pantry. Some of these recipes are easy as can be and some require at least some time and patience, but not a lot of skill or culinary expertise.

With a busy work schedule and cold nights, a crab casserole lets you assemble everything and then pop it in the oven while you do something else. Some of these recipes aren't exactly one-pot meals because they involve a sauté or some other preliminary step or steps, but the cooking mess is pretty manageable. They also let you use a combination of crabmeat types or just the special so you have a slight feeling of decadence without breaking your food budget.

When you're using cheese in these recipes, you may want to try a variety of cheeses to see which complement the crabmeat and which overwhelm it. And with today's hunger for bacon, you can try adding it to the casserole or soufflé.

*Crabs not in season*

# Stuffed Portobello Mushrooms

I find the combination of mushrooms and crabmeat warms my tummy as soon as I start buying the ingredients and lasts until the last crumb has been indelicately and impolitely picked up on my fingertip. This recipe is a little more ingredient intensive than the stuffed mushrooms in the appetizer section. It's a little more dramatic, too.

1 lb. crab claw meat
½ cup minced onion
2 tbsp. parsley, minced
1 tbsp. garlic powder
½ tsp. dry mustard
½ tsp. kosher salt
Pepper, to taste
Juice of 1 lemon
1 tbsp. Worcestershire sauce
½ cup mayonnaise
¼ cup heavy cream
1 cup seasoned dry stuffing mix, pounded into coarse crumbs, or other crumbs of your choice
1½ oz. Italian cheese (or cheeses), shredded
½ cup Parmesan cheese, grated, divided
¼ cup white wine
8 Portobello mushroom caps
Olive oil
Dash paprika

Gently combine first 12 ingredients plus 1 cup of the shredded cheese and ¼ cup of the Parmesan cheese. Refrigerate overnight or at least 3 to 4 hours.

Preheat oven to 400 degrees. Line a large baking pan with foil. Pour wine into the pan.

Clean mushrooms with a brush, not water. Clean out the dark gills. Brush olive oil on the tops of the caps and place top side down in baking pan. Place crab mixture into the caps and top with the remaining cheeses. Sprinkle with paprika. Bake 25 to 30 minutes or until golden and caps are cooked.

# Crab-Stuffed Poblano Pepper with Roasted Tomato Sauce and Cilantro Oil

Todd and Ellen Kassoff Gray are co-owners of Equinox restaurant and Todd Gray's Watershed, both in Washington, D.C. Opened in 1999, Equinox has become a hugely popular and highly regarded restaurant dedicated to supporting seasonality and sustainable farming. The Grays work with farmers within a one-hundred-mile radius of the city and have their own line of Black Angus beef from Warrenton, Virginia.

As this was being written, on a cold, snowy January day, where their menu features seasonal blood oranges, golden pineapple, Meyer lemon, and crispy Rappahannock oysters, the Grays changed seasons with this crab meets chili dish. As mild as poblano peppers are (remove the seeds), if they're still too hot for you, feel free to substitute a green, red, or yellow bell pepper.

1 lb. crabmeat, jumbo lump, cleaned
¼ cup mayonnaise
¼ tsp. cayenne pepper
Salt and pepper, to taste
6 poblano peppers, grilled and peeled
2 cups fresh tomato sauce or prepared gourmet variety
¼ cup cilantro oil or other type of gourmet herb oil

Preheat oven to 350 degrees

In a mixing bowl, combine crabmeat, mayonnaise, cayenne pepper, and salt and pepper; mix well. Stuff peppers with crabmeat and place on lightly oiled baking dish. Bake peppers for 20 to 30 minutes until heated through. Warm sauce in a small saucepot over medium heat. Heat 6 salad plates. Spoon tomato sauce in center of plate, top with stuffed poblano pepper, and drizzle with herb oil. Serve immediately.

# Flounder Stuffed with Crab

One of my fondest memories is of my parents driving from Silver Spring out to where I lived in Bowie to take me and my daughters to dinner while my (then) husband worked late. I think our favorite restaurant was Busch's Chesapeake Inn, owned by Robert Randolph Busch, Sr. It was an extraordinary place that had more culinary stars than you can imagine, which was notable because it was totally out in the boonies. It became more isolated when Route 50 was widened and the only access was off a ramp and Busch's Frontage Road. Mr. Busch, named to the Maryland Restaurant Association's Hall of Honor, retired in 1999, closing the restaurant after fifty-three years of operation. He died in December 2001, at the age of seventy-five.

My favorite dish was the Flounder Stuffed with Crabmeat. This recipe seems to fit my memory. I hope it's close. You can use the Mornay sauce or eliminate it.

Serves 3

1 cup onion, minced
½ cup celery, minced
½ cup parsley, minced
2 cloves garlic, minced
3 tbsp. butter
1 tbsp. all-purpose flour
½ cup milk
½ cup white wine
2 cups crabmeat
Salt and pepper, to taste
6 flounder filets
Mornay Sauce (see recipe)
Dash paprika

Sauté the onion, celery, parsley, and garlic in butter until tender. Stir in the flour. Gradually add the milk and wine and cook until slightly thickened. Remove from the heat and gently stir in the crabmeat. Season to taste.

Arrange 3 filets on a jellyroll pan that's been sprayed with vegetable oil, and spoon about 1 cup of the crab mixture on each. Cut the remaining filets in half, lengthwise, and drape around the crab mixture. Cover with the Mornay Sauce and sprinkle with paprika.

Bake at 425 degrees for 15 minutes or until the fish is done.

## Mornay Sauce

2 tbsp. butter
¼ cup all-purpose flour
2 cups milk
Salt and white pepper, to taste
2 tbsp. heavy cream
¼ cup Swiss or Gruyere cheese, shredded

Melt butter and flour to create a roux and cook for about 1 minute, until pale yellow, not brown. Gradually add the milk, stirring constantly. Cook until thickened and bubbly. Add the salt and pepper to taste. Add the cream and then add the cheese and stir until it's melted. Add a little more milk or cream if sauce is too thick.

# Seafood Newburg

This is a variation of the very rich Lobster Newburg, which reportedly was created by a sea captain, Ben Wenberg, who showed it to the manager at Delmonico's restaurant in Manhattan. The chef, Charles Ranhofer, refined it and put it on the menu as Lobster a la Wenberg. The dish was taken off the menu after a dispute between Wenberg and Delmonico, yet the clientele clamored for its return. It reappeared as Lobster a la Newburg. You can add mushrooms or change the amount of seafood you use, including lobster. Remember, the crabmeat is already cooked, so wait until late in the cooking process to add it.

Serves 8

½ lb. scallops
1 pt. oysters
6 tbsp. butter
⅓ cup flour
½ tbsp. salt
¼ tsp. cayenne
¼ tsp. nutmeg
1 cup milk
1 cup whipping cream
3 egg yolks
½ cup dry sherry
1 lb. crabmeat
½ lb. shrimp, cooked

Sauté the scallops and oysters in 2 tbsp. of butter. Remove from the butter and set aside. Stir the flour and spices into the butter. Cook, stirring constantly, until it's bubbly and then add the milk and cream. Stir until the sauce thickens and then remove from the heat.

Beat the egg yolks in a small bowl and stir in the sherry. Put ½ cup of the hot sauce mix into the egg yolks and gently mix. Add the egg mixture into the rest of the sauce. Add the crab, shrimp, scallops, and oysters and warm until the crab is heated. Serve over rice or fettuccini noodles.

# Seafood Supreme

This is a variation on Lobster Newburg that eliminates the eggs and the need to temper the dish. It's also baked, so you have time to work on the rest of your menu. Feel free to vary the amount of lobster, crab, shrimp, scallops, and oysters.

Serves 8

½ cup butter
3 tbsp. flour
1 qt. light cream
Salt and pepper, to taste
¼ tsp. paprika
½ lb. lobster
½ lb. crabmeat
1 lb. shrimp
1 lb. scallops
1 pt. oysters
¼ cup Parmesan cheese

Make a roux with the butter and flour and then add the cream and stir. Cook for a few minutes, just until the flour is no longer raw. Add seasonings to taste.

Add the lobster, crab, shrimp, scallops, and oysters. Place in a buttered casserole dish and bake at 350 degrees for about 20 to 30 minutes or until bubbly. Top with cheese and return to the oven for 5 minutes or until brown.

# Soft-Shell Crab Imperial

As mentioned a few times in this book, doing anything to the basic crab could be considered overkill, and yet we continue to add goodness to deliciousness. This recipe from the Virginia Marine Products Board, of Newport News, Virginia, is a perfect example.

Serves 6

¼ cup butter
2 tbsp. flour
1 cup milk
2 tsp. prepared mustard
½ tsp. salt
⅛-¼ tsp. red pepper
1 lb. crabmeat, back fin
12 soft-shell crabs, cleaned
3 tbsp. butter, melted
1 cup fresh buttered breadcrumbs

Melt ¼ cup butter in a saucepan. Add the flour and blend over low heat, stirring constantly, for 3 to 5 minutes. Slowly stir in milk. Cook and stir until thickened. Blend in the mustard, salt, and red pepper. Gently fold in crabmeat.

Prick the legs and claws of each soft-shell crab with the tines of a fork to prevent popping. Place the crabs, bottom side up, on a broiling rack 3 inches from the heat. Brush with melted butter. Broil 4 to 5 minutes. Turn over and brush again with butter. Broil 4 to 5 minutes.

Place about 3 tbsp. of the crab mixture on top of each broiled crab. Sprinkle with buttered crumbs. At serving time, bake in a preheated 350-degree oven for 15 minutes.

# Stuffed Soft-Shell Crabs

This very easy recipe comes from the Maryland Seafood Marketing and Aquaculture Development Program, and it's a great change from stuffing with a Crab Imperial mix. Hey, it's soft-shell crabs and back fin crabmeat; what more could you want?

Serves 6

12 medium soft-shell crabs, cleaned
1 lb. back fin crabmeat
½ cup (1 stick) butter or margarine

Dry crabs with paper towels. Remove all cartilage from crabmeat. Place crabs in shallow baking pan. Remove top shell from crabs and stuff each crab cavity with about 3 tbsp. crabmeat. Replace top shell. Melt butter and pour evenly over crabs. Bake at 400 degrees until shells turn red and crabs brown slightly, about 15 minutes.

*Crab feast (Courtesy Shoreline Seafood, Gambrills, Maryland)*

# Soft-Shell Crabs with Lime Butter, Bay Spice, and Baby Spinach

One of the finest locavores in the area, Chef Todd C. Gray, of Equinox restaurant, breathes new life into historic recipes using Virginia Piedmont ingredients.

One ingredient you're sure to notice is the New Bay spice. This is a special blend that Chef Todd mixes, and it can be purchased at Market Salamander in Middleburg, Virginia, or through their Web site. You may, of course, substitute Old Bay.

Serves 6

3 tbsp. New Bay spice, divided
3 cups Wondra brand flour
6 soft-shell crabs, cleaned
2 tbsp. clarified butter plus ⅓ cup
Juice of 2 limes
1 tbsp. parsley, chopped
3 cups baby spinach, cleaned and wilted in olive oil
3 shallots, sliced and sautéed until softened
¼ cup red peppers, roasted and cut into ¼ inch strips
Salt and pepper, to taste

Heat a large sauté pan to medium-high temperature. Combine 2 tbsp. bay spice with flour. Dip crabs in flour, shaking off excess. Add 2 tbsp. clarified butter to pan; lightly sauté crabs 3 minutes per side. Remove and keep warm.

Add 1 tbsp. bay Spice to pan and allow heat to release aroma. Add remaining butter and heat until it turns brown. Add lime juice and parsley, keep warm.

Combine spinach with shallots and peppers. Adjust seasoning and heat in a warm oven. Heat 6 large plates and spoon a small amount of spinach mixture in the center of each plate. Top with soft-shell crab and spoon New Bay Spice Lime Butter over and around crabs. Serve immediately.

*(Courtesy Equinox restaurant)*

# Fried Soft-Shell Crabs

This is the recipe to use when you want to fry those soft shells. Replace the pancake flour with your coating of choice, and, if you wish, shred some lettuce in a hoagie bun and insert fried crab.

1 cup pancake flour
1 tsp. Old Bay seasoning
12 soft-shell crabs, cleaned
Fat for frying

Mix flour and seafood seasoning. Dredge crabs in the flour mixture to coat well. In large electric frying pan, heat about ½ inch of cooking oil to 375 degrees. Add crabs and turn heat down to 350 degrees. Cook crabs about 5 minutes on each side (coating will brown).

# Soft-Shell Crab Stack

As if soft-shell crabs aren't decadent and succulent by themselves, the Virginia Marine Products Board has come up with this majestic mountain of mouth-watering goodness. The recipe includes instructions for making Hollandaise, but you're allowed to buy a jar from the grocery store. This dish is good for dinner and probably even better for breakfast or brunch, akin to eggs Benedict, without the eggs (but you could add a fried egg to the stack).

Serves 4 to 8

8 soft-shell crabs, cleaned
3 tbsp. butter
4 English muffins, split, toasted, and buttered
8 slices cooked ham, ⅛ to ¼ inch thick, the size of an English muffin
8 slices ripe tomato
Hollandaise Sauce (see recipe)

Prick the legs and claws of each crab with fork tines to prevent popping. Sauté in butter about 4 minutes on each side over moderate heat.
Place toasted English muffin halves on plates. Top each with a ham slice, then tomato, and then cooked crab. Spoon about 3 tbsp. Hollandaise Sauce over each.
Serve immediately.

## Hollandaise Sauce

4 egg yolks
3 tbsp. lemon juice
¼ tsp. salt
¼ tsp. white pepper
⅛ tsp. cayenne pepper
1 cup butter, melted

Place egg yolks, lemon juice, and seasonings in a blender. At medium speed, very slowly add 1 cup butter that has been melted to bubbling but not browned. Blend an additional 10 to 12 seconds until sauce is thickened and smooth.

# Crab au Gratin

Here's an example of the dish being the exception to the rule for it does combine fish (crab) and cheese. Oh well. We just have to ignore the rule sometimes. This gratin can be prepared ahead and heated a few minutes before serving. Plus, once you have the crabmeat, you probably have everything on hand, so it doesn't require a major shopping trip.

4 tbsp. butter or margarine, divided
½ green pepper, minced
½ onion, chopped
3 tbsp. flour
2 cups milk
2 cups crabmeat
½ tsp. salt
Dash ground nutmeg
½ cup shredded cheese, American or a variety of your favorite cheeses
4 oz. breadcrumbs

Melt 2 tbsp. butter and add pepper and onion and cook 5 minutes. Add the flour and milk, then crabmeat, salt, and nutmeg. Cook 10 minutes. Pour into a shallow, buttered baking dish or individual ramekins or baking dishes.

Either melt 1 tbsp. butter and combine with the breadcrumbs until they're evenly buttered (be careful not to brown) or place both into a blender until the breadcrumbs and butter are evenly distributed. The later avoids cooking the breadcrumbs before you put them in the oven.

Sprinkle each dish with shredded cheese and buttered breadcrumbs. Bake at 350 degrees, about 10 to 15 minutes or until the cheese is brown.

# Crabmeat au Gratin

This is another version of a crab au gratin dish, again with items probably already in your pantry.

½ cup green onions or scallions, chopped
½ cup fresh parsley, chopped
4 tbsp. butter
3 tbsp. all-purpose flour
3 tbsp. sherry
1 ½ cups half-and-half
1 lb. crabmeat
8 oz. cheese, shredded, American, mixed, or pizza blend
Salt and pepper, to taste

Sauté onions and parsley in the butter. Remove from heat and stir in the flour. Remove from the pan. Add the sherry to the pan to deglaze and then gradually add the half-and-half and cook over low heat, stirring constantly until it thickens. Return the onions, parsley, and flour mixture to the sauce. Add salt and pepper to taste.

Gently distribute the crabmeat evenly in a 2-qt. baking dish. Pour the sauce over crabmeat. Cover with cheese. Bake at 375 degrees for 10 to 15 minutes or until the cheese is bubbly.

# Deviled-Crab-Filled Crêpes

This is a buttery, creamy mixture that can be used to fill crêpes or baked in ramekins.

1 cup cracker crumbs
1 tbsp. lemon juice
1 tsp. Worcestershire sauce
½ cup onion, minced
Dash of hot sauce
Dash of cayenne pepper
1 tsp. dry mustard
¼ cup parsley, chopped
⅔ cup butter, melted
¼ cup evaporated milk
1 lb. crabmeat
Salt and pepper, to taste
1 package or 24 crêpes (see recipe)
½ cup cracker crumbs, for topping
Butter, for topping

In a bowl, mix the first ten ingredients together and gently incorporate the crabmeat into the mix. Season with salt and pepper. You can refrigerate the ingredients overnight, and then bring to room temperature before filling the crêpes.

Spoon some mixture into each crêpe and roll. Sprinkle cracker crumbs and a small dab of butter on top of each crêpe. If you prefer, put the mixture into a baking dish and sprinkle the cracker crumbs and pieces of butter on top. Heat in a 350-degree oven for 10 to 15 minutes.

## Crêpe Batter

3 eggs
2 cups flour
¼ cup melted butter
¼ tsp. salt
2 cups milk

Combine ingredients in a blender and blend for about 1 minute. Scrape down the sides with a rubber spatula and blend until smooth, about 15 to 20 seconds. Refrigerate batter at least 1 hour so the flour has a chance to rest. Cook on upside-down crêpe griddle or in traditional pan.

*Fresh crabs sign*

# Deviled Crab

Serves 6

1 cup breadcrumbs
Juice of 1 lemon
1 tsp. Worcestershire sauce
2 tbsp. butter, melted
4 tbsp. mayonnaise
½ tsp. salt, to taste
½ tsp. pepper, to taste
4 hard-cooked eggs, chopped
1 lb. crabmeat, back fin

Mix all ingredients together except the eggs and crabmeat. Gently add the chopped eggs and the crabmeat. Place in individual shells or baking dishes and bake in 350-degree oven for 20 to 25 minutes.

*Seafood store*

# Deviled Crab, Two

This variation starts with sautéed onions and a roux and changes things up a little by adding chopped herbs into the breadcrumb topping.

2 tbsp. onion, chopped
3 tbsp. butter, melted, divided
2 tbsp. flour
¾ cup milk
1 tbsp. lemon juice
1½ tsp. dry mustard
Dash pepper, to taste
Dash cayenne pepper, to taste
1 tsp. Worcestershire sauce
1 egg, beaten
1 tbsp. parsley, chopped
1 lb. crabmeat, special
¼ cup dry breadcrumbs
1 tbsp. thyme, chopped
1 tbsp. dill, chopped

Cook onion in 2 tbsp. butter until tender. Blend in flour. Add milk gradually and cook until thick, stirring constantly. Add lemon juice and seasonings. Add the Worcestershire sauce to the egg, gradually, and then pour the egg mix into the sauce, stirring constantly. Add parsley and crabmeat and blend gently, but well. Divide into 6 greased individual ramekins or 5-oz. custard cups.

Combine 1 tbsp. butter, crumbs, and any herbs you're using. Sprinkle over the top of each shell. Bake in a moderate oven, 350 degrees, for 20 to 25 minutes or until brown.

# Deviled Crab, Three

Start with some soaking wet bread that's not quite dripping with liquid. The richer the liquid (e.g., heavy cream), the richer the dish. Use water instead of dairy if you prefer.

4 slices bread
¼ cup cream, milk, or water
1 lb. crabmeat, back fin
½ cup cracker meal
2 tbsp. creamy salad dressing (French, thousand island, ranch, etc.)
1 tsp. prepared mustard
½ tsp. Tabasco sauce
1 tsp. Worcestershire sauce
4 stalks celery, chopped
½ green pepper, chopped
1 small onion, chopped
Salt and pepper, to taste
Dash paprika

Soak the bread in the cream briefly, just long enough to be damp, and then break into small pieces. Mix in the crabmeat, cracker meal, and salad dressing. Add the mustard, Tabasco, and Worcestershire. Add the celery, green pepper, and onion to the crabmeat mixture. Add salt and pepper to taste. Place in crab shells, ramekins, or 2-qt. dish. Sprinkle with paprika. Place on baking sheet and bake for 45 minutes in a 325-degree oven.

# Crab Pot Pie

This great comfort recipe takes a little preparation time and then you let it do its thing. You will be flooded with all the memories of eating pot pie when you were growing up with the surprise of crabmeat instead of chicken or turkey. The silkiness of the creamy sauce combined with the sweet crabmeat makes this recipe a favorite for many generations.

Serves 4 to 6

1 cup onion, chopped
1 cup celery, chopped
1 cup carrots, chopped
3 oz. melted butter
½ cup flour
¼ cup sherry
2 cups of chicken stock
1½ cups heavy cream
¼ cup frozen or fresh peas
1½ lb. crabmeat
1 tsp. Old Bay seasoning
½ tsp. thyme
1 pie dough

Heat oven to 400 degrees.
Sauté the onion, celery, and carrots in butter until they are soft but still slightly firm. Next, add the flour to the vegetables and cook for about 3 minutes. Add the sherry, chicken stock, and cream and stir until smooth. Add the peas and crab and simmer for about 4 minutes. Add the seasonings and simmer for 1 minute longer.
Roll the pie dough to fit the bottom and sides of a 2-qt. baking dish (or something similar). The dough should hang over the sides by at least 2 inches and be about ¼ inch thick.
Put the mixture in the baking dish and cover with the pie dough. Seal the edges and prick with fork tines or slice steam vents. Bake for 30 minutes until the crust is golden brown.

# Fish en Papillote

As we eat with our eyes, presentation goes a long way to making something taste sensational. This is true when the presentation is *en papillote* or "in parchment." This recipe comes from Susan Trombetti, a local if there ever was one. She enjoys experimenting with this recipe and has used garlic, scallops, and shrimp, although she thinks the scallops and crab work best together.

You can make this dish with the filet flat or rolled up around the vegetables and crabmeat or rolled around the vegetables with the crabmeat on top. Professor Mark Ainsworth of the Culinary Institute of America says you should roll it skin side out (with the skin removed) to enable it to cook evenly. The crabmeat doesn't overcook, he says, because of steam in the pouch. Choose the vegetables of your choice as long as they're cut to about the same size to assure the same doneness. Although preparation time is a little long, truly the toughest part of the meal is to secure the parchment paper around the dish and then opening it at the dinner table. As Julia Child would say, *"Bon appétit!"*

4 salmon filets (about 6-7 oz.), skin removed
4 sheets of parchment paper or aluminum foil
8 oz. green beans, julienne slice
8 oz. carrots, julienne slice
 Juice of 1 lime
Juice of 1 lemon, then cut into quarters or eighths
Dill or fresh parsley
4 oz. crabmeat, lump
1 tbsp. butter
Salt and fresh pepper, to taste
2 oz. white wine

Preheat the oven to 400 degrees.

Place one filet (skin side down, if the skin were still on the filet) in the center of a 10-inch square of parchment paper. Place the vegetables over the filet and sprinkle with the lime and lemon juices and dill. Gently place the crabmeat on top of the filet and vegetables. Place dots of butter over vegetables and filet and season with salt and pepper. Splash with wine. Either leave flat or roll the filet, jellyroll style, from the tail to the head so it circles the vegetables. Bring 2 opposite sides of parchment square together and fold to seal, forming an envelope.

Or take the parchment paper and fold one corner to the other, to form a triangle over the fish. Starting with one end fold the paper, about 1 inch at a time, along the edge so it creates a hem or sealed bag. Continue to the end and make the last fold over itself. The object is to seal in all the ingredients.

Bake for 10 to 12 minutes or until the parchment paper is brown. If you used foil, carefully open/slit the foil, taking care to avoid the steam and plate. If you used the parchment, plate it and serve to your guests with the bag intact. Provide them with a knife or scissors to open the bag. Alternatively, you can slit the bag open before you make the presentation.

# Potato-Chip-Encrusted Striped Bass with Sautéed Back Fin Crab, Papaya, and Fresh Pineapple

The Virginia Marine Products Board provided this dish of fish and crab with a Pacific fusion touch by Chef Chuck Sass of Mahi Mah's Seafood Restaurant and Sushi Saloon, in Virginia Beach, Virginia. Use a little imagination and you may hear and feel the vibe from the outdoor patio and stage, the ocean waves breaking over the beach, and the gentle breeze going through your hair.

4 6 oz. skinless striped bass filets
1 oz. flour
3 oz. buttermilk
2 cups salted, plain potato chips, crushed
4 oz. clarified butter
2 tbsp. chopped shallot
½ cup diced fresh papaya
½ cup diced fresh pineapple
1 oz. dry white wine
1 oz. freshly squeezed orange juice
2 tbsp. cilantro
½ lb. crabmeat, back fin
3 tbsp. whole butter, soft
Salt, to taste
White pepper, to taste

Dredge the striped bass filets in flour, then buttermilk, and finally the crushed potato chips. Heat the clarified butter in a large skillet and brown the filets on both sides, being careful not to burn the potato chips. When brown, place on a baking pan and bake in a 350-degree oven for approximately 5 minutes or until filets are cooked.

Pour off excess grease from the skillet and add shallot, papaya, and pineapple. Cook for 1 minute on medium heat, and deglaze pan with the white wine and the orange juice. Add cilantro and gently fold the crab and the whole butter into the sauce. Season with salt and pepper. Remove the filets from the oven and pour the sauce over the fish.

*(Courtesy Virginia Marine Products Board)*

# Basque Crab Casserole

Carolyn McHugh is president and CEO of the Calvert County (Maryland) Chamber of Commerce, so you can understand when she says she entertains a lot. This recipe, she says, "came from a friend, but, like all the recipes that I use, I made it mine by changing some things and adding some of my own favorite ingredients. I'm constantly on the prowl for new ideas and interesting ways to serve a number of people without breaking the hostess budget."

Carolyn notes, "With the addition of a mild white fish to a smaller amount of crabmeat, this delicious casserole easily serves four relatively inexpensively without diminishing the luscious crab flavor."

Serves 4

4 tbsp. canola oil
1 onion, finely chopped
2 tbsp. green pepper, finely minced
¼ cup celery, finely chopped
8 oz. fresh or frozen crabmeat, picked clean
1 lb. tilapia or other mild white fish, cooked and flaked
½ cup sherry
1 cup tomato sauce or sofrito
4 oz. cream cheese, room temperature
½ cup artichoke hearts, chopped
5 oz. dry white wine
Pinch of cayenne
Salt and pepper, to taste
1 tbsp. chopped parsley
Bottled clam juice or milk (optional)
4 tbsp. Japanese panko (plain, unseasoned breadcrumbs may be substituted)
1 tbsp. butter

In a frying pan, heat the oil and sauté the onion, green pepper, and celery until softened. Add the crabmeat and flaked, cooked fish to the pan. In a small pan, heat sherry and tomato sauce together over a low flame. When slightly reduced, add the cream cheese. Then add the tomato sauce/cream cheese mixture, artichoke hearts, wine, cayenne, salt, pepper, and parsley to the crabmeat mixture. Cook for

15 minutes, adding a little bottled clam juice or milk to thin the sauce if needed. Spoon the mixture into 4 oiled ramekins. Sprinkle the tops with panko and dot with butter. Place under broiler flame just until the tops are browned, about 5 minutes.

*Empty crab bushels*

# Beef Wellington and Crab in Phyllo Dough

My friends and I have been celebrating New Year's Eve together for many years. We take turns hosting and preparing a meal fit for royalty. I was planning to make a modified Beef Wellington one year and realized that most of us like our meat on the rare side, but one prefered it akin to shoe leather. I knew it was going to be difficult to please everyone. I resolved the issue by making individual Wellingtons. They were a hit and so is this transformation of a beef dish to one that incorporates crab.

I recommend you buy puff-pastry sheets from the freezer section of the grocery store rather than try to make the pastry yourself. Remember to remove the package from the freezer about 6 hours prior to assembly and cooking or you will be making phyllo cookie crisps.

4 sheets of phyllo dough, kept moist and cool
1 stick butter, melted and clarified
2 tsp. olive oil
½ lb. mixed mushrooms, chopped (shiitake, button, cremini, oyster, etc.)
3 tbsp. dry red wine
3 tbsp. green onions, chopped
¼ tsp. dried thyme leaves, crushed
Salt and pepper, to taste
4 small beef tenderloin steaks, cut 1 inch thick
½ lb. crabmeat, back fin

Spread melted butter on the first phyllo sheet (keep a damp towel over the other sheets until used). Stack a second sheet on top of it and butter. Continue with the other 2 sheets. Fold the 4 sheets in half so you have a rough square. Slice the 8 sheets into equal-sized quarters. Cover with damp cloth until you're ready to assemble.

Preheat oven to 425 degrees.

Heat oil in large nonstick skillet over medium heat until hot. Add mushrooms; cook and stir 5 minutes or until tender. Add wine and cook 2 to 3 minutes or until liquid is evaporated. Stir in green onions, thyme, and salt and pepper to taste. Remove from skillet and cool completely.

Heat the same skillet over medium heat until hot. Place steaks in skillet and cook 3 minutes, turning once. (Steaks will be partially cooked. Do not overcook.) Cook filets a little longer if you prefer meat well done. Season with salt and pepper, as desired.

Place about 2 tbsp. of the mushroom mixture in the center of each phyllo sheet. Put a piece of meat on top of the mushroom mixture. Divide the crab and place some on top of each steak. Bring all 4 corners of each phyllo stack together; twist tightly to close. Lightly spray each bundle with cooking spray or brush with melted butter. Place on greased baking sheet. Immediately bake in oven for 10 minutes or until golden brown. Let stand 5 minutes before serving.

*Crab tools*

# Sautéed Soft-Shell Crabs with Julienne of Cucumber, Country Ham, and Shiitake Mushroom with Herb Butter

Virginia Marine Products Board provided this recipe from former chef and part owner Marcel Desaulniers of the Trellis in Williamsburg, Virginia. He is also director emeritus of the Culinary Institute of America. The "Guru of Ganache" broke away from his love of chocolate long enough to create this masterpiece.

4 tbsp. lemon juice
12 soft-shell blue crabs, cleaned
Salt and pepper, to taste
½ cup all-purpose flour, seasoned with salt and pepper
6 tbsp. unsalted butter, divided
6 cucumbers, peeled, seeded, and julienned
½ lb. shiitake mushrooms, sliced
¼ lb. country ham, julienned
Herb Butter Sauce (see recipe)

Sprinkle lemon juice over crabs and season with salt and pepper. Cover and refrigerate until ready to cook.

Pat crabs dry with paper towel. Dust with flour using up to ½ cup, as necessary. Heat 2 tbsp. butter in a large sauté pan. When butter is hot, carefully sauté crabs, shell side first, until golden brown on both sides. Remove crabs from pan and hold warm.

In a clean sauté pan, sauté cucumbers in 2 tbsp. of butter. In a separate pan, sauté mushrooms and ham in 2 tbsp. of butter. When all the ingredients are hot, assemble the dish.

To assemble, place cucumbers on individual dinner plates. Portion 2 to 3 tbsp. of Herb Butter Sauce over each plate of cucumbers. Set soft-shell crabs on top of cucumbers (2 per plate). Finish by spooning shiitake and ham mixture over the crabs. Serve immediately.

## Herb Butter Sauce

1 tbsp. all-purpose flour
1 tbsp. butter, softened
¼ cup fish stock
¼ cup white wine
¼ cup unsalted butter
1 tbsp. fresh herbs, chopped (chives, parsley, lemon thyme)
Salt and pepper, to taste

Combine flour and softened butter until mixture is smooth. In a small pan, heat fish stock and white wine to a boil. Quickly whip butter and flour mixture into boiling hot stock and wine. Remove the pan from heat. Whip ¼ cup butter into thickened mixture 1 tbsp. at a time. When all the butter has been incorporated into the sauce, add the chopped herbs. Season with salt and pepper to taste. Hold away from heat until served.

# Crab Enchiladas

Create your own Tex-Mex variation with a Bay touch.

Serves 3 to 4

1 medium onion, chopped
2 tbsp. cooking oil
2 tbsp. flour
2½ cups milk
1 7-oz. can green chiles, drained and chopped
½ lb. Monterey Jack cheese, grated and divided
1 dozen corn tortillas
½ lb. crabmeat

Sauté onion in oil. Add flour, milk, chiles, and 4 oz. of cheese and cook until medium thick.

Heat oil in a separate pan and dip each tortilla in hot oil for a few seconds until it becomes limp.

Put 1 tbsp. each of the cheese, crabmeat, and melted sauce on each tortilla, then roll it up. Place tortillas on a shallow baking dish and cover with the rest of the sauce. Bake, uncovered, at 250 degrees for 10 to 15 minutes until the sauce is bubbly.

*Trays of crabs*

# Crab Cobbler

½ cup butter or margarine
½ cup green pepper, chopped
½ cup onion, chopped
½ cup sifted all-purpose flour
1 tsp. dry mustard
1 cup milk
1 cup mild Cheddar cheese, shredded
2 tsp. Worcestershire Sauce
Salt and pepper, to taste
1 cup crabmeat (any type or mixture)
1½ cups tomatoes (drain if using canned tomatoes)
Cheese Biscuit dough (see recipe)

Melt butter or margarine in top of double boiler. Add pepper and onion and cook over boiling water for about 10 minutes or until tender. Blend in flour, mustard, milk, and shredded cheese. Cook, stirring constantly, until cheese is melted and mixture is thick. Add Worcestershire, salt and pepper, crabmeat, and tomatoes. Blend thoroughly and pour into a 2-qt. casserole. Drop Cheese Biscuit dough by rounded teaspoonfuls on top of the crabmeat mixture. Bake at 450 degrees for about 15 to 20 minutes or until brown and bubbly.

## Cheese Biscuit Dough

1 cup sifted all-purpose flour
2 tsp. baking powder
½ tsp. salt
¼ cup Cheddar cheese, shredded
2 tbsp. butter or margarine
½ cup milk

Sift flour, baking powder, and salt. Add cheese. Cut in butter or margarine until mixture is like coarse meal. Add milk. Mix only until all the flour is dampened.

# Rod 'N' Reel Restaurant Crab Imperial

Since 1946, the Rod 'n' Reel restaurant has been an institution in Chesapeake Beach. This was more than a decade after the demise of the Chesapeake Beach Railroad, and they've seen plenty of changes in those sixty years, from the days of slot machines (legalized in 1948) to the explosion of condos. Although it seems very far away, Chesapeake Beach is about thirty miles from Washington, D.C., so commuting is tempting to some.

Four generations later, Joyce Stinnett Baki grew up in the family-owned restaurant and provides this recipe for Crab Imperial from Wesley Donovan.

Serves 40

1 gallon Hellmann's mayonnaise
1 cup red bell pepper, minced
1½ cups green bell pepper, minced
10 lb. crabmeat, colossal lump
Imperial Glaze (see recipe)

Serves 4

12 oz. Hellmann's mayonnaise
1 oz. red bell pepper, minced
1.5 oz green bell pepper, minced
1 lb. crabmeat, lump
Imperial Glaze (see recipe)

Mix the mayonnaise and peppers together and then gently fold in the lump crabmeat. Place in casserole dishes (individual or large family style). Place in a preheated 400-degree oven for approximately 15 minutes. Remove from oven and top with Imperial Glaze. Return dish to oven and bake until the glaze is brown and bubbly.

# Imperial Glaze

Serves 40

1 qt. Hellmann's mayonnaise
8 egg yolks
1 tbsp. Old Bay seasoning
2 tbsp. fresh lemon juice

Serves 4

3 oz. Hellmann's mayonnaise
1 egg yolk
Dash Old Bay seasoning
Squeeze of fresh lemon juice

Mix all ingredients.

*Fresh crabs sign*

# Maryland Crab Imperial Casserole

As mentioned elsewhere, David DeBoy created the perennial "Crabs for Christmas" melody while his wife has created some tasty recipes. When entertaining a larger crowd, Joellen will double this recipe and bake and serve in a 13x9-inch casserole dish. David notes that this is Joellen's recipe, "so it's not written as funny as mine."

½ green pepper, finely chopped
½ red pepper, finely chopped
6 tbsp. butter
1 lb. crabmeat
2 heaping tbsp. mayonnaise
1-2 dashes Worcestershire sauce
1 tsp. mustard
½ tsp. salt
¼ cup Cheddar cheese, grated
¼ cup breadcrumbs

Sauté peppers in butter. Mix in the remaining ingredients except the cheese and breadcrumbs and toss lightly. Place in a 1½-qt. casserole dish. Sprinkle top with cheese and breadcrumbs. Bake at 400 degrees for 20 minutes.

# Senator Mikulski's Crab Imperial

1 lb. crabmeat, back fin
Dash of salt, to taste
Cayenne pepper, to taste
1 green pepper, diced
2 eggs, well beaten (set aside 2 tbsp.)
5 tbsp. mayonnaise, divided
1 tbsp. chopped onion

Preheat oven to 350 degrees.

Combine the crabmeat with salt, cayenne, green pepper, beaten eggs (minus the 2 tbsp.), 4 tbsp. of mayonnaise, and chopped onion. Fill 6 decorative serving shells or ramekins with the crabmeat mixture. Add 1 tbsp. mayonnaise to the remaining 2 tbsp. of beaten egg and pour over each filled shell. Sprinkle with cayenne pepper. Bake at 350 degrees for about 30 minutes.

# Crab Soufflé

Frank Skalitza is a treasure of a chef who's hidden away at the Canvasback Restaurant in downtown Cambridge, Maryland. A classically trained chef and longtime resident of the Ocean City, Maryland, area, Frank created a repertoire of dishes he could fix in ten minutes for a previous restaurant, which were sold for $10.00. This crab soufflé is part of his collection. This will make about four servings, but it could be cut to serve one perfectly.

Serves 4

4 eggs, separated
1 oz. dry sherry
1 tsp. Dijon mustard
1 oz. chives, minced
Dash of Worcestershire sauce
1 tsp. Old Bay seasoning
4 oz. mayonnaise
1 lb. fresh crabmeat, back fin
Butter

Preheat oven to 350 degrees.
Blend together the egg yolks, sherry, mustard, chives, Worcestershire sauce, Old Bay, and mayonnaise, and then gently fold in the crabmeat.
In a separate bowl whip the egg whites to soft peaks and then fold into the crab mixture. Lightly butter 4 6-oz. ramekins and fill with mixture. Bake for 10 to 12 minutes or until internal temperature reaches 130 degrees.

# Deviled Crab Soufflé

This soufflé combines the best of the drama of a soufflé and the flavors of a deviled crab mix. When cooking a soufflé or other recipe in a pan filled with water (*bain-marie*), put the dish or dishes in the baking pan and then fill with hot water. This eliminates having to carry a heavy pan filled with water and, even more important, reduces the chance that you'll be burned by the hot water sloshing around in and out of the pan.

2 tbsp. butter
¼ cup flour
½ tsp. salt
½ tsp. dry mustard
1 cup milk
3 egg yolks, beaten
2 tbsp. fresh parsley, chopped
2 tsp. finely chopped onion
1 tbsp. lemon juice
1 lb. crabmeat
3 egg whites, beaten to soft peaks
Butter or oil

Melt the butter and blend in the flour and seasonings. Add milk gradually and cook until thick and smooth, stirring constantly.

Stir a little of the hot sauce into the beaten egg yolks and then add the egg mix to the sauce, stirring constantly.

Add parsley, onion, lemon juice, and crabmeat. Fold in egg white. Place in a well-greased 1½-qt. soufflé dish or baking pan. Place the dish in a pan of hot water. Bake in a moderate oven, 350 degrees, for 1 hour or until soufflé is firm in the center. Serve immediately.

# Crab and Cheese Bake

This one-dish meal whips up in no time and then takes care of itself in the oven while you're working on other parts of your meal, entertaining friends, or accomplishing the thousands of things that must be done. You can try switching the kind of cheese and the type of bread used to satisfy the palates of those who will be dining or according to what's in stock.

1 lb. crabmeat
2 tbsp. onion, chopped
2 tsp. mustard
1 cup celery, chopped
¾ cup mayonnaise
16 slices American cheese
16 slices white sandwich bread, crusts removed
4 eggs, beaten
1 tsp. Worcestershire sauce
2 cups milk
1½ tsp. salt

Mix first 5 ingredients.
Butter or oil a 12x9 pan. Place 8 slices of bread in the pan. Cover each with a slice of cheese. Spread crab mixture over bread and cheese. Cover with remaining bread and cheese.
Mix egg, Worcestershire sauce, milk, and salt and pour over bread stacks. Chill for at least 8 hours.
Preheat oven to 325 degrees and bake for 1 hour.

# Crab Ravioli in Creamy Tomato Sauce

This recipe and the following one take advantage of packaged wonton wrappers to speed the preparation work. This one uses a dried tomato soup mix. You can mix and match the flavors you like.

8 oz. crabmeat
½ cup ricotta cheese
1 tsp. Italian seasoning
1 package wonton wrappers
1 egg, beaten
1 cup milk
1 package dried creamy tomato sauce mix, or use canned tomato soup and ¼ cup milk
2 tbsp. butter
Fresh parsley, chopped

In a small bowl, gently blend the crabmeat, cheese, and seasoning. Let it sit for a few minutes to blend the flavors.

With a wonton wrapper facing you like a baseball field, put about ½ tsp. of the crab mixture in the center, brush the edges with the beaten egg (water will do), and fold the wrapper from home plate to second base to form a triangle. Press any air out of the wonton package and seal the edges together. Place on a floured surface.

Bring a large pot of kosher salt and water to boil.

Add the milk to a medium pot and heat to just before the boiling stage. Combine the dry sauce mix and milk and bring to a boil. Reduce the heat and simmer, stirring frequently, for about 3 minutes. Reduce the heat to keep the sauce warm, but not bubbling.

Depending on the size of the pot with the boiling water, drop 6 to 8 ravioli into the pot, being sure not to crowd the pillows. Cook for 2 minutes or until the ravioli rise to the top. Remove the ravioli, making sure the liquid drains off, and add to the sauce.

Sprinkle with the parsley.

# Crab Ravioli in a Creamy Shallot Sauce

8 oz. crabmeat
1 cup ricotta cheese
1 tsp. oregano
1 package wonton wrappers
1 egg, beaten
1 tbsp. olive oil
1 shallot, finely chopped
½ cup dry white wine
1 cup fish or chicken broth
½ pt. heavy cream
1 tbsp. parsley, chopped
Salt and pepper, to taste
1 tbsp. salt
Parmigiano-Reggiano cheese

In a small bowl, gently blend the crabmeat, cheese, and oregano. Let it sit for a few minutes so the flavors blend.

With a wonton wrapper facing you like a baseball field, put about ½ tsp. of the crab mixture in the center, brush the edges with the beaten egg (water will do), and fold the wrapper from home plate to second base to form a triangle. Press any air out of the wonton package and seal the edges together. Place on a floured surface.

Bring a large pot of water and kosher salt to boil.

Heat the olive oil in a skillet and sauté shallots until tender. Add the wine, broth, and cream, simmering until thickened. Add the parsley, salt, and pepper to taste. Let it sit over a low temperature.

Depending on the size of the pot with the boiling water, drop 6 to 8 ravioli into the pot, being sure not to crowd the pillows. Cook for 2 minutes or until the ravioli rise to the top. Remove the ravioli, making sure the liquid drains off, and add them to the sauce. Sprinkle with Parmigiano-Reggiano cheese if desired.

# Manicotti with Crab Sauce

This is one easy dish that tastes like you spent the entire day or weekend cooking.

Serves 4

2 8-oz. cans tomato sauce of your choice
8 oz. crabmeat
8 manicotti shells
1 cup Parmesan cheese, grated, divided
2 eggs, beaten
2 cups large-curd cottage cheese
2 tbsp. parsley, chopped

Heat the oven to 350 degrees.
Warm the tomato sauce and add the crabmeat. Spread ⅓ of the mixture in the bottom of a 1½-qt. baking dish.
Cook the shells according to package directions, drain, and rinse (unless instructed not to).
Mix ½ cup of the Parmesan cheese, the eggs, cottage cheese, and parsley. Fill the shells with this mixture and arrange on top of the sauce. Spoon the remaining sauce over the filled shells. Bake, covered, for 25 minutes or until the shells are hot. Sprinkle with the remaining cheese.

*Crab sign*

# Crabmeat and Spaghetti

Beth Rubin is an Annapolis-based writer and editor who says she spends almost as much time in her kitchen as at her computer.

She says, "I got hooked on my Nana Helen's lobster and spaghetti at family dinners. Not an in-the-kitchen grandmother, she copped the recipe from her sister's Irish cook! At home my mother would make this for company and I would snitch some when she wasn't looking. Packing the recipe in my trousseau, along with monogrammed towels and lacy garter belts, I continued the tradition with my family and guests. As the price of canned lobster meat skyrocketed, I began using crabmeat and found I like it better. Now my children, grandchildren, and guests lap it up. What's not to like?"

About this recipe, she suggests you "have your cholesterol level checked before you even prepare it and definitely before eating."

Serves 3 to 4

5 oz. extra-sharp Cheddar cheese*
12 tbsp. (1½ sticks) butter, plus enough to grease casserole dish
10 oz. ketchup
1 tbsp. Worcestershire sauce
¼ tsp. dry mustard
Salt and pepper, to taste
½ lb. spaghetti**
12 oz. crabmeat, lump, cleaned
4 oz. dry breadcrumbs

Preheat oven to 375 degrees.

Grate the cheese and set aside.

In small saucepan, melt butter over very low heat. As it melts stir in the ketchup until well blended. Add Worcestershire sauce, dry mustard, salt, and pepper and stir. Remove from the heat.

Cook the spaghetti according to package instructions for "al dente" (slightly underdone). Do not let it go to mush. Transfer to colander, run under cold water, and drain well.

Assemble the ingredients in a buttered ovenproof casserole in this order: half the spaghetti, half the crabmeat, half the sauce. Repeat. Sprinkle the cheese and breadcrumbs on top.

Bake for 30 to 40 minutes.

Serve with a mixed green salad and crusty bread. A little goes a long way.

*Grate the cheese yourself. Do not buy the prepackaged shredded cheese.

**Spaghetti works best in this recipe. Avoid using vermicelli, capellini, or angel hair.

*Beth Rubin (Photograph by Mac Bogert)*

# Crab Macaroni and Cheese

Macaroni and cheese has gone upscale with a version in Santa Monica containing white truffle and selling for $95.00. Adding crab to this comfort food makes it more exciting and doesn't set you back more than a C-note (you have to tip the server and you'll probably have a beverage and you should have a salad). Feel free to change the cheeses so the final dish tastes the way you remember it, or try a more daring cheese that will add a bite to the meal. You will need about two cups of shredded cheese, some Parmesan cheese, and some ricotta as well. You can change the liquid portions as long as you end up with about 3½ cups of milk or milk-like products. Yes, you may use broth, but it won't be as creamy.

12 oz. shells or elbow noodles
1½ cups whole milk
2 cups heavy cream or half-and-half, or increase the amount of whole milk
2 tsp. all-purpose flour
½ tsp. salt
¼ tsp. freshly ground black pepper
1 cup bleu cheese, grated
¾ cup Cheddar cheese, grated
½ cup Parmesan cheese, grated
¼ cup mozzarella cheese, grated or shredded
¾ cup ricotta cheese
10 oz. crabmeat
4 tbsp. breadcrumbs
2 tbsp. unsalted butter

Preheat the oven to 450 degrees. Spray a 13x9-inch glass baking dish with cooking spray.

Cook the noodles in a large pot of boiling salted water until tender but still firm to the bite, stirring frequently, depending on the package instructions. Drain well and do not rinse unless the package directions say you should.

Combine the milk, cream, flour, salt, and pepper in a large pot and set on medium heat. Stir in the cheeses. Add the noodles, swirling through the cheese sauce until all the pasta is covered. Gently fold in the crabmeat.

Pour the mac and cheese into the prepared baking dish and sprinkle with breadcrumbs. Place dabs of butter on top.

Bake for 15 to 20 minutes or until the mixture is brown and bubbly. Let it rest for a few minutes before serving.

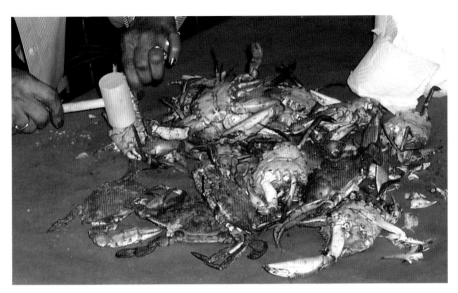

*Digging in at a crab feast*

# Dreamy Crab Noodle Casserole

Trish Weaver, of Dream Weaver Events and Catering in Prince Frederick, Maryland, says, "One cold winter day when we were wishing it was summertime we put together a dish to remind us of summer days. It is our way to have 'summertime' comfort food in the middle of the winter. Sometimes we crumble up left over crab cakes (not that we have leftovers very often) on top instead of the breadcrumbs. It goes well as a lunch special in the café with a garden salad and warm bread . . . and for catering. We have offered it with a seafood raw bar, with fried chicken, or as a lunch for an office." This version will make enough for six to eight people as a main course, more if it's used as a dip or spread or, as Trish uses it, part of a seafood buffet or raw bar.

Serves 6 to 8

1 each red, yellow, green pepper, chopped
1 red onion, chopped
½ lb. mushrooms, cleaned, stemmed, and sliced
1 package egg noodles, prepared per package directions
1 cup butter
1 cup all-purpose flour
1 cup milk
¼ lb. each of Cheddar, Monterey Jack, and Colby cheeses, shredded in processor
Cajun seasoning, to taste
Minced garlic, to taste
1 lb. crabmeat
Breadcrumbs

In a pan, sauté peppers, onion, and mushrooms in butter until just soft. Set aside.

Cook noodles according to package directions. While they're cooking, make a roux with equal parts butter and flour. Add milk to thin to creamy consistency. Add three-quarters of the shredded cheeses to make a smooth cheese sauce, stirring until melted and blended well. Add Cajun seasoning and garlic to taste.

Pour sauce over cooked egg noodles and blend in sautéed vegetables. Place in a shallow baking dish. Fold in the crabmeat lightly so as not to break up the crab. Top with remaining cheeses and breadcrumbs and bake at 350 degrees until hot and melted all the way through. Serve immediately.

*(Courtesy Dream Weaver Events and Catering)*

*Soft & Salty Seafood*

# Desserts

I was a little surprised that I was unable to find a huge selection of desserts that featured crabmeat. However, I did find some references to crab-flavored ice cream. According to the Food Network, which cites the *Guinness Book of World Records*, the ice cream parlor with the most flavors of ice cream, including cream of crab, is in Merida, Venezuela. Reportedly, Heladeria Coromoto, set in the Andean hills, has a repertoire of 860 flavors. Another shop in Hokkaido, Japan (in the northern part of the country where they have a lot of seafood), supposedly carries crab-flavored ice cream, too. Ask for Kani Aisu (crab ice cream).

Closer to home, Chip Hearn, of Peppers Ice Cream in Rehoboth, Delaware, tried making crab-flavored ice cream. Should you want to experiment, he recommends using lump back fin in large chunks and gently folding it in at the refrigerator-temperature stage instead of at the beginning or at the freezer-temperature stage. He paid $50 for two pounds of crabmeat and his experiments ran the gamut. Finally settling on butter-flavored ice cream as a base, he figured he'd have to charge $10 a cone. Instead, his customers had a chance to taste it at regular price and enjoyed it, but don't look for it on any kind of a regular basis.

"Yuck" is the usual response I receive when I talk about crab ice cream. However, Heston Blumenthal of the *Guardian* newspaper said in 2003, "It is not just taste that determines whether we like a particular ingredient or dish. All of the senses play a part, as does memory. But play around with those influences and you can override all preconceptions." His article cites Dr. Charles Spence, an experimental psychologist at Oxford University (the one in England) who said, "If you tell people they are going to eat crab ice cream, they'll recoil. But tell them that it's frozen crab bisque and they'll eat it without a fuss. That's because we associate the title 'ice cream' with desserts."

Victor Barlow, of Scottish Highland Creamery in Oxford, Maryland, occasionally makes Old Bay sorbet and on even fewer occasions makes Old Bay and crab sorbet as part of his selection of more than six hundred flavors. His wife, Susan, says they found that ice cream and crab didn't work well together, but the seasoning with the sorbet does. The recipes are a well-guarded secret and I'm not about to spoil their fun. Victor grew up in Edinburgh, Scotland, and occasionally you'll find him in his Scottish finery with his ice cream cart in the park, aboard the Oxford-Bellevue Ferry. You just never know where he'll appear.

# Crab Crème Brûlée

Marcus Bradley, chef at the Irish Eyes Pub in Lewes, Delaware (and Rehoboth and Milton), came to our rescue with a dessert using crabmeat. As mentioned in the sauce section, where you can find his recipe for Chesapeake Aioli, Marcus has spent time cooking in North Carolina; Washington, D.C.; and New Orleans and has "been working with crabs for many years."

You will need a mini blowtorch to caramelize the sugar.

Serves 4

2 cups heavy cream
¼ cup white sugar
Pinch of kosher salt
Juice of 1 lemon
3 oz. crabmeat, back fin
3 egg yolks
2 eggs
16 pieces lump crabmeat
1 tsp. sugar

Preheat oven to 300 degrees.

Combine the cream, sugar, salt, lemon, and back fin crabmeat. Bring to a slow boil, stirring occasionally, for 4 to 5 minutes. Set aside to cool slightly.

Beat the yolks smooth and slowly add the whole eggs one at a time. Add the warm cream mixture to the egg mixture, a little at a time, stirring constantly, until all the cream is incorporated. Pour the mixture into 4 6-oz. ramekins. Drop 4 pieces of lump crabmeat into each ramekin.

Place the ramekins in a baking dish and fill the dish with hot water until it reaches halfway up the ramekins. Cover the pan loosely with foil and bake for 25 to 30 minutes until the custard is just set. Chill the ramekins in the refrigerator for 2 to 3 hours.

Before serving sprinkle sugar on top and torch!

# Crab Cookies for Christmas

In the almost-there category is David DeBoy, who mentioned that he makes crab-shaped cookies for Christmas. You may remember him from earlier in this book as the man who created the song "Crabs for Christmas."

You will need a crab cookie cutter to make these cookies.

## Cookie Dough

¾ lb. (3 sticks) unsalted butter, room temperature
1 cup sugar
1 tsp. pure vanilla extract
3½ cups all-purpose flour, plus extra for dusting
¼ tsp. salt

## Cookie Frosting

6 tbsp. water
1 lb. confectioners' sugar
1 tsp. almond extract
Red food coloring
Yellow food coloring

## Decorating Options

Black decorating gel
Red or orange gum drops
Red sugar crystals
Silver decors balls

## Cookie Dough

Let the butter come to room temperature. If you're impatient like me, throw the butter into the microwave for 18 seconds.

Using an electric mixer, slowly blend the butter and sugar together in a tall bowl. I use a tall bowl because if I don't I tend to decorate my walls with batter when I raise the electric beaters. Once the butter and sugar are fluffy, add the vanilla, and mix a little more. Try not to decorate your walls.

Sift the flour and salt and add into the sugar and butter conglomeration. Fire up the electric mixer on low and blend it into a doughy mess.

*David DeBoy, recorded the hit song "Crabs for Christmas." (Courtesy David DeBoy)*

Dust a clean, dry surface with flour and drop the cookie dough onto it with a resounding "plop"! Mold it with your hands into a shape roughly resembling a turtle shell. Cover it in plastic and refrigerate for at least 30 minutes.

Preheat the oven to 350 degrees.

Roll the dough out on the aforementioned flour-dusted surface to a thickness of about ½ inch. Cut out crab shapes with the flour-dusted crab cookie cutter and place the cookies on an ungreased baking sheet.

"Now, the trick to keeping the claws from breaking off of the soon-to-be-baked cookies," says David, "is to gently push the raw, doughy claws up against the shell body of the crab. If the claws are just hanging out there vulnerable and defenseless, they have a tendency to snap off before reaching the mouth of any potential crab cookie eater.

"Stop fooling around and put the cookies into the oven. Bake for 20 to 25 minutes, watching for the edges to brown a bit. Remove the tray and transfer the crabby cookies to a wire rack until they cool to room temperature . . . unless you keep your room temperature over 100 degrees," he continues.

*(Photograph by David DeBoy)*

Cookie Frosting

In a large bowl fold 6 tbsp. of water into the confectioners' sugar until you can form soft peaks resembling the Big Horn Mountains of Wyoming. Mix in the almond extract.

Add the red food dye a drop at a time to transform the icing to an acceptable steamed-crab color. If you prefer more orange than red, add some yellow food coloring along with the red.

Ice the cookies with a soft spatula or spoonula or your fingers. Or spoon some icing into a food storage bag, clip off a corner, and pipe the icing at your discretion. You can also forego the icing and sprinkle red or orange sugar crystals on your crabs. Get creative! Use the black icing gel to draw outlines of the crab or draw in their happy face. Always make them smile! Use tiny silver decors for eyes.

If you want to impress, roll out orange or red gumdrops with a rolling pin. (Cover the rolling pin with plastic wrap and wet it with water to keep the gumdrops from sticking to the pin.) Once you've flattened the gumdrops into a sheet, use the crab cookie cutter to cut out a crab shape and stick it to the top of a cookie using some icing for glue.

Now, find them a good home in the mouths of your friends and family.

# Index

4th & Swift's Summer Sweet Corn Soup with Lump Crab, Chives, and Old Bay, 190
1972 Clean Water Act, 37

## A

Adam's Ribs, 77
Admiral Fell Inn, 100
Adryon, 150, 174
Ainsworth, Mark, 70
allergy, 59
American Farm Bureau, 38
Annapolis Maritime Museum, 28
Annapolis Rotary Club, 76
Anne Arundel Community College, 100
Annual Soft-Shell Spring Fair, 74
Aunt Stell's Crab Cakes, 154

## B

Baki, Joyce Stinnett, 134
Basic Crab Salad, 166
Basque Crab Casserole, 232
Beach, Gary, 106, 119, 186
Beef Wellington and Crab in Phyllo Dough, 234
Bradley, Marcus, 204, 259
Brooks, Jack, 25

## C

Calvert Marine Museum, 29
Calypso Crab Cakes, 136
Cambridge, Maryland, 31, 75
Captain John Smith Chesapeake National Historic Trail, 14
Captain Salem Avery House, 29
Cardin, Benjamin L., 37
Carnival Pride, 35
Carroll, Nicholas, 153
Chesapeake Aioli, 204
Chesapeake Bay Blue Crab Cakes, 144
Chesapeake Bay Bridge, 35
Chesapeake Bay Crab Cake with Bloody Mary Salsa, 156
Chesapeake Bay Foundation, 14, 38
Chesapeake Bay Maritime Museum, 29
Chesapeake Bay Program, 16
Chesapeake Crab and Beer Festival, 75
Chesapeake Crab Cake, 142
Chesapeake Crab Quiche, 88
Chessie, 46
Clayton Co., J. M., 25
Clyde Bernard "Bernie" Fowler, 36-39
Clyde's Crab Salad Tower, 168
Clyde's Restaurant Group, 168
Clyde's Restaurant Group Maryland Crab Soup, 180
Coast Day, 135-36, 140, 142
Cold Crab and Avocado Soup, 181

Cold Crab Claw Soup, 188
Conservation Fund, 14
Crab and Apple Beignets, 90
Crab and Cheese Bake, 246
Crab and Corn Muffins, 82
Crab and Oysters Bubb-afeller, 104
Crab and Shrimp Ceviche with Watermelon Relish, 126
Crab and Tomato Soup, 191
Crab Asparagus Soup, 200
Crab au Gratin, 220
Crab Bisque, 184
Crab Cake Guy, 131
Crab Cakes and Polenta with Spicy Sweet Potatoes and Old Bay Crema, 150
Crab Cheese Spirals, 96
Crab Cobbler, 239
Crab Cocktail, 116
Crab Cookies for Christmas, 260
Crab Crème Brûlée, 259
Crab Cutlet (Crab Cake Variation), 160
Crab Deviled Eggs, 101
Crab Dip, 106
Crab Dip, Creamier Style, 113
Crab Dip for a Group, 112
Crab Dip, Slightly Spicy, 107
Crab Dip Two, 109
Crab Dip with Mushrooms, 108
Crab Drop, 40
Crab Dumplings, 124
Crab Enchiladas, 238
Crab Macaroni and Cheese, 252
Crabmeat and Spaghetti, 250
Crabmeat au Gratin, 221
Crab Meltaways, 117
Crab Muffins, 84-85
Crab Po' Boys, 177
Crab Pot Pie, 227
Crab Pretzel, 114
Crab Pretzel Two, 115
Crab Ravioli in a Creamy Shallot Sauce, 248
Crab Ravioli in Creamy Tomato Sauce, 247
Crab Remoulade, 205
Crab Rolls, 97
Crabs for Christmas, 117
*Crabs for Christmas*, 45
Crab Soufflé, 244
Crab Spinach Salad, 172
Crab Spread, 121
Crab Spring Rolls, 122
Crab-Stuffed Poblano Pepper with Roasted Tomato Sauce and Cilantro Oil, 209
Crab Tacos, 173
Crab Tacos with Old Bay Slaw and Charred Corn and Avocado Salsa, 174
Crab Tarts, 95
Crab Wontons in Creamy Shallot Sauce, 120
Cream of Crab Soup, 192
Crêpe Batter, 222
Crisfield, Maryland, 30, 75
Crooks, Patrick, 146
croqueta de jaiba, 132
Culinary Institute of America, 70
Cummings, Elijah, 37

# D

Davidson, Stuart, 180

DeBoy, David, 45, 117, 242, 260
DeBoy, Joellen, 46, 117, 242
derelict crab traps, 16
Deviled Crab, 224
Deviled Crab, Breakfast Style, 89
Deviled-Crab-Filled Crêpes, 222
Deviled Crab Soufflé, 245
Deviled Crab, Three, 226
Deviled Crab, Two, 225
*Dirty Jobs*, 27
Discovery Channel, 20
Dorchester County, 23
Douglass, Jackie Leatherbury, 40
Dreamy Crab Noodle Casserole, 254
Dungeness crab, 20

### E

Easton, Maryland, 36, 40-41
Easy Peasy Cream of Crab Soup, 194
Edwards, Jerry, 182
egg-bearing females, 15
*Enchantment of the Seas*, 35

### F

Fab Blue Crab Taboo, 111
Fish en Papillote, 228
Flounder Stuffed with Crab, 210
Founding Farmers Crab Devil-ish Eggs, 102
Founding Farmers Devil-ish Egg Salad, 103
Founding Farmers Hot Maryland Crab Dip, 110
Founding Farmers Louie Dressing, 103
Founding Farmers Mini Crab Roll, 176
Fournier, Elizabeth, 59, 206
Freeman, Heather, 145
Freytag, Pamela, 149
Friedman, Jimmy, 45
Fried Soft-Shell Crabs, 218
Fruchthendler, Saul, 81
Fugere, Mary L., 164

### G

ghost traps, 16-17
Gibson Island Crab Cakes, 149
Gill, Vanessa, 194
Golden Crab Bisque, 182
Gordon, Barry, 154
Gordon, Renee S., 154
Governor J. Millard Tawes Historical Museum and Ward Brothers Workshop, 30
Gray, Ellen Kassoff, 209
Gray, Todd C., 209, 216
Green Onion Dipping Sauce, 206

### H

H-2B visa, 24
Haggard, Joe, 124
Hampton Roads Naval Museum, 31
Hampton (Virginia) Crab Salad, 171
Harbor Magic Hotel, 100
Harbor Magic Maryland Crab Cake, 143
Hardesty, Brent, 45

Havre de Grace Maritime Museum, 29
Herb Butter Sauce, 237
Historic London Town and Garden, 29
Historic St. Mary's City, 30
Holguin, Stephanie, 111
Hollandaise Sauce, 219
Hood, Richard and Suzanne, 41
Hooper Strait Lighthouse, 29
horseshoe crab, 21, 32
Hutt, Mike, 144

## I

Imperial Glaze, 241
inactive crabbing commercial licenses, 23
International Human Rights Law Clinic, 24

## J

Jimmies, 47, 49
Joe's Crabbies, 118
John Smith Chesapeake Trail, 14
Johnson, George, 136
Joyce, Joe, 142
Jumbo Lump Crab Bruschetta, 119
Jumbo Lump Crab Cakes with Spicy Remoulade, 158

## K

Kaine, Tim, 37
king crab, 20
Koco's Pub, 132

## L

Legal Sea Foods Crab Cakes, 145
Lemon Crab Soup with Vegetables, 199
Leonardtown, 75
Lewes, Delaware, 135
Licurgo, Jenna, 167

## M

Manicotti with Crab Sauce, 249
Mariner's Museum, 31
Maryland Crab Imperial Casserole, 242
Maryland Department of Natural Resources, 17, 22-23, 51
Maryland Lady Crab Cakes, 133
Maryland Seafood Festival, 76
Maryland Seafood Marketing and Aquaculture Development Program, 53, 201
Mason, John, 184
McDonnell, Bob, 16
McHugh, Carolyn, 232
Meyer, Thomas, "Tom," 192
Mikulski, Barbara, 24, 148
Mornay Sauce, 211
Museum of Chincoteague Island, 32
Mushrooms Chesapeake, 100

## N

Nappo, Al, 102, 110, 176
Narrows Restaurant, 132
National Association of Farm Bureaus, 38
National Geographic Society, 14

National Harbor, Maryland, 75
National Hard Crab Derby & Fair, 75
Nauticus, 31-32
Nicholas Carroll's Spicy Crab Cakes, 153

## O

O'Malley, Martin, 16, 34, 37
Open Gates Farm Bed & Breakfast Crab Salad, 167
Oriole Park at Camden Yards, 100
Overmiller, Shannon, 195

## P

Pan-Seared Blue Crab Cakes with Pepper and Onion Jam, 138
Parkhill, Charles, 140
Peppers Ice Cream, 257
Pink Floyd, 93
Plummer, Kirk G., 205
Helen Delich Bentley Port of Baltimore, 34
Potato-Chip-Encrusted Striped Bass with Sautéed Back Fin Crab, Papaya and Fresh Pineapple, 230
Pumpkin Crab Soup, 189

## Q

Quick Crab Bisque, 185
Quik Pik, 26

## R

Reedville Fishermen's Museum, 33
Richardson Maritime Museum, 31
River and Harbor Act of 1970, 34
Roasted Corn and Crab Chowder, 186
Rod 'n' Reel restaurant, 240
Rod 'n' Reel Restaurant Crab Imperial, 240
Rowe, Mike, 27
Roy's Baltimore Crab Cake Recipe, 146
Rubin, Beth, 250
Ruiz, Steven M., 138

## S

Sandy Point State Park, Maryland, 76
Sargent, William, 21
Sautéed Soft-Shell Crabs with Julienne of Cucumber, Country Ham, and Shiitake Mushroom with Herb Butter, 236
Schmidt, Jimmy, 156
SCK Maryland Style Crab Cakes, 161
Seafood Newburg, 212
Seafood Supreme, 213
Senator Barb's (Mikulski) Favorite Crab Cakes, 148
Senator Mikulski's Crab Imperial, 243
She-Crab Soup, 195
She-Crab Soup with Sherry and Whipping Cream, 198
Shoreline Seafood, 50, 65, 77
Skalitza, Frank, 244
Smith, Captain John, 14

snow crab, 20
Soft-Shell Crab Imperial, 214
soft-shell crabs, 54, 62
Soft-Shell Crab Stack, 219
sooks, 15, 47
South City Kitchen, 86, 126
South City Kitchen Crab Hash, 86
South City Kitchen Old Charleston She-Crab Soup, 196
Spicy Crab Soup, 201
Spider Sushi Maki Roll, 127
Stallman, Bob, 38
Starkey, Keith, 142
*State of Play*, 42
Stinnett, Elizabeth, 134
Stinnett Family Crab Cakes, 134
St. Mary's College of Maryland, 30
St. Mary's Crab Festival, 75
stone crabs, 19
Storm, Donald, 50
Storm, Mike, 42, 65
Stuffed Mushrooms, 98
Stuffed Portobello Mushrooms, 208
Stuffed Soft-Shell Crabs, 215
Sullivan, Bryan, 143
Sussex County Low Country Crab Cakes with Crabanero Remoulade Sauce, 140
swamp dogs, 50
Swift, Jay, 190

## T

Taste of Cambridge Crab Cook-off, 75
Timbuktu Crab Cakes, 132
Todd, Travis, 104
Trombetti, Susan, 228
Tunks, Jeff, 158

## U

Ulbrich, Chip, 86, 90, 126, 161, 196
United States Department of Commerce, 19
University of Delaware, 135
USA American Blue Crab, 23

## V

Van Hollen, Chris, 37
Vaughan, Nancy, 149
Vintage Crab Salad, 164
Virginia Institute of Marine Science, 17-18
Virginia Marine Products Board, 19, 88, 214, 219, 230, 236
Virginia Marine Resources Commission, 17-18, 23

## W

Weave, Trish, 254
Wesley, Donovan, 134
West Point Crab Carnival, 76
West Point, Virginia, 76
William Preston Lane, Jr. Memorial Bridge, 35
Wine Coach, the, 93
Winnifred and August F. Crabtree, 32
WMPT, 45

## Y

Yamaguchi, Roy, 146